Silvanus P. Thompson

Light Visible and Invisible

A Series of Lectures Delivered at the Royal Institution of Great Britain, at Christmas,

1896

Silvanus P. Thompson

Light Visible and Invisible

A Series of Lectures Delivered at the Royal Institution of Great Britain, at Christmas, 1896

ISBN/EAN: 9783337069513

Printed in Europe, USA, Canada, Australia, Japan

Cover: Foto ©berggeist007 / pixelio.de

More available books at **www.hansebooks.com**

LIGHT, VISIBLE AND INVISIBLE

LIGHT

VISIBLE AND INVISIBLE

A SERIES OF LECTURES
DELIVERED AT THE ROYAL INSTITUTION OF
GREAT BRITAIN, AT CHRISTMAS, 1896

BY

SILVANUS P. THOMPSON,

D.Sc., F.R.S., M.R.I.

PRINCIPAL OF, AND PROFESSOR OF PHYSICS IN, THE CITY AND GUILDS
TECHNICAL COLLEGE, FINSBURY, LONDON

New York

THE MACMILLAN COMPANY

LONDON: MACMILLAN AND CO., Limited

1897

INTRODUCTION

Two things are expected of a lecturer who undertakes a course of Christmas lectures at the Royal Institution. In the first place his discourses must be illustrated to the utmost extent by experiments. In the second, however simple the language in which scientific facts and principles are described, every discourse must sound at least some note of modernity, must reflect some wave of recent progress in science.

So in undertaking a course of lectures in Optics the lecturer in the present instance ventured to proceed on certain lines which may, perhaps, seem strange to the sedate student whose knowledge of optics has been acquired on the narrower basis of the orthodox textbook. The ideas developed in the first lecture arose from the conviction that the time-honoured method of teaching geometrical optics — a method in which the wave-nature of light is steadily ignored — is fundamentally wrong. For the sake of students and teachers of optics he has added to Lecture I. an Appendix, in which the newer ideas are further developed. Other Appendices have been added to the later Lectures, with the aim of filling up some of the gaps left in the subjects as treated in the lecture theatre.

Now that the electromagnetic nature of all light-waves has been fully demonstrated, no apology is needed for bringing into the fifth Lecture a few of the experimental points upon which that demonstration rests. That these fundamental points can be given without any great complication of either thought or language is in itself the strongest argument for making that demonstration an essential feature at an early stage in the teaching of the science.

Many of the ideas which must be grasped, for example that of the polarisation of light, are popularly supposed to be extremely difficult; whereas the difficulty lies not in the ideas themselves so much as in the language in which they are generally set forth. In an experience lasting over a good many years, the author has found that the main points in the phenomena of polarisation are quite easily grasped by persons of ordinary intelligence—even by children—provided they are presented in a modern way devoid of · pedantic terms, and illustrated by appropriate models. A similar remark would equally apply to other parts of optics, such as interference and diffraction, which are barely alluded to in the present lectures. Many branches are necessarily omitted altogether from so brief a course: amongst them the entire subject of spectrum analysis, the construction and theory of optical instruments, and the greater part of the subject of colour vision. No attempt was made to include these topics, and no apology is needed for their omission. Whatever value these discourses may possess must depend upon the things they include, not upon those which they do not.

LONDON, *January* 1897.

CONTENTS

LECTURE III

POLARISATION OF LIGHT

LECTURE IV

THE INVISIBLE SPECTRUM (ULTRA-VIOLET PART)

The spectrum stretches invisibly in both directions beyond the visible part—Below the red end are the invisible longer waves that will warm bodies instead of illuminating them—These are called the calorific or *infra-red* waves. Beyond the violet end of the visible spectrum are the invisible shorter waves that

LECTURE V

THE INVISIBLE SPECTRUM (INFRA-RED PART)

LECTURE VI

RÖNTGEN LIGHT

LECTURE I

LIGHTS AND SHADOWS

LIGHT, as is known both from astronomical observations
and from experiments made with optical apparatus, travels
at a speed far exceeding that of the swiftest motion of any
material thing. Try to think of the swiftest thing on the
face of the earth. An express train at full speed, perhaps, occurs to you. How far will it go while you
count up to ten? Counting distinctly I take just over
$5\frac{1}{2}$ seconds. In that time an express train would have
travelled 500 feet! Yet a rifle-bullet would have gone
farther. There is something that goes quicker than any
actual moving thing. A sound travels faster. In the
same time a sound would travel a mile. Do you say
that a sound is only a movement in the air, a mere aerial

B

wave? That is quite true. Sound consists of waves, or rather of successions of waves in the air. None of you who may have listened to the delightful lectures of Professor M'Kendrick in this theatre last Christmas will have forgotten that; or how he used the phonograph to record the actual mechanical movements impressed by those air-waves as they beat against the sensitive surface of the tympanum.

But this Christmas we have to deal with waves of a different kind—waves of light instead of waves of sound —and though we are still dealing with waves, yet they are waves of quite a different sort, as we shall see.

In the first place, they travel very much faster than waves of sound in the air. During that $5\frac{1}{2}$ seconds, while an express train could go 500 feet, or while a sound would travel a mile, light would travel a million miles! A million miles! How shall I get you to think of that distance? An express train going 60 miles an hour would take $16,666\frac{2}{3}$ hours, which is the same thing as 694 days 10 hours 40 minutes. Suppose you were now—29th December 1896, 3 o'clock—to jump into an express train, and that it went on and on, not only all day and all night, but all through next year, day after day, and all through the year after next, month after month, until November, and that it did not stop till 24th November 1898 at 20 minutes before 2 o'clock in the morning; by that time—nearly two years—you would have travelled just a million miles. But the space that an express train takes a year and eleven months to travel, light travels in $5\frac{1}{2}$ seconds—just while you count ten!

And not only are the waves of light different from those of sound in their speed—they are different in size. As compared with sound-waves they are very minute ripples. The invisible waves of sound are of various sizes, their lengths differing with the pitch of the sound. The middle c' of the pianoforte has a wave-length of about 4 feet 3 inches, while the shrill notes that you can sing may be only a few inches long. A shrill whistle makes invisible ripples about half an inch long in the air. But the waves of light are far smaller. The very largest waves of all amongst the different kinds of visible light —the red waves—are so small that you could pack 39,000 of them side by side in the breadth of one inch ! And the waves of other colours are all smaller. How am I to make you grasp the smallness of these wavelets ? What is the shortest thing you can think of ? The thickness of a pin ? Well, if a pin is only a hundredth part of an inch thick it is still 390 times as broad as a ripple of red light. The thickness of a human hair ? Well, if a hair is only a thousandth part of an inch thick it is still 39 times as big as the size of a wave of red light.

Now, from the facts that waves of light travel so fast, and are so very minute, there follow some very important consequences. One consequence is that the to-and-fro motions of these little ripples are so excessively rapid— millions of millions of times in a second—that there is no possible way of measuring their frequency : we can only calculate it. Another consequence is that it is very difficult to demonstrate that they really *are* waves. While a third consequence of their being so small is that, unlike big waves, they don't spread much round the edges of obstacles.

You have doubtless all often watched the waves on the sea, and the ripples on a pond, and know how when the waves or the ripples in their travelling strike against an obstacle, such as a rock or a post, they are parted by it, pass by it, and run round to meet behind it. But when waves of light meet an obstacle of any ordinary size they don't run round and meet on the other side of it—on the contrary, the obstacle casts a shadow behind it. If the waves of light crept round into the space behind the obstacle, that space would not be a dark shadow.

Well, but that is a question after all of the relative sizes of the obstacle and of the waves. Sea waves may meet behind a rock or a post, because the rock or the post may not be much larger than the wave-length.[1] But if you think of a big stone breakwater—much bigger in its length than the wave-length of the waves,—you know that there may be quite still water behind it; in that sense it casts a shadow. So again with sound-waves; ordinary objects are not infinitely bigger than the size of ordinary sound-waves. The consequence is that the sound-waves in passing them will spread into the space behind the obstacle. Sounds don't usually cast sharp acoustic shadows. If a band of musicians is playing in front of a house, you don't find, if you go round to the

[1] Note that the scientific term "wave-length" means the length from the crest of one wave to the crest of the next. This, on the sea, may be 50 feet or more. In the case of ripples on a pond, it may be but an inch or two. Many people would call it the breadth of the waves rather than the length.

back of the house, that all sound is cut off. The sounds spread round into the space behind. But if you notice carefully you will observe that while the house does not cut off the big waves of the drum or the trombone, it does perceptibly cut off the smaller waves of the flute or the piccolo. And Lord Rayleigh has often shown in this theatre how the still smaller sound-waves of excessively shrill whistles spread still less into the space behind obstacles. You get sharp shadows when the waves are very small compared with the size of the obstacle.

Perhaps you will then tell me that if this argument is correct, you ought not, even with light-waves, to get sharp shadows if you use as obstacles very narrow obstacles, such as needles or hairs. Well, though perhaps you never heard it, that is exactly what is found to be the case. The shadow of a needle or a hair, when light from a single point or a single narrow slit is allowed to fall upon it, is found not to be a hard black shadow. On the contrary, the edges of the shadow are found to be curiously fringed, and there is light right in the very middle of the shadow caused by the waves passing by it, spreading into the space behind and meeting there.

However, all this is introductory to the subject of shadows in general. If we don't take special precautions, or use very minute objects to cast shadows, we shall not observe any of these curious effects. The ordinary shadows cast by a bright light proceeding from any luminous point are sharp-edged; in fact, the waves, in ordinary cases, act as though they did not spread into the shadows, but travelled simply in straight lines.

Let me try to illustrate the general principle of the

FIG. 1.

travelling of ripples by use of a shallow tank [1] of water,

[1] Ripple-tanks for illustrating the propagation of waves have long been known. Small tanks were used at various times by Professor Tyndall. See also Professor Poynting, F.R.S., in *Nature*, 29th May 1884, p. 119.

on the surface of which I can produce ripples at will. An electric lamp placed underneath it throws up shadows of the ripples upon a slanting translucent screen, and you can see for yourselves how the ripples spread from the centre of disturbance in concentric circles, each circle enlarging, and the ripples following one after another at regular distances apart. That distance is what we call the "wave-length."

If I use the tip of my finger to produce a disturbance, the ripples travel outward in all directions at an equal speed. Each wave-front is therefore a circle. If, however, I use to produce the disturbance a straight wooden ruler, it will set up straight wavelets that follow one another in parallel ranks. These we may describe as plane waves, as distinguished from curved ones. Notice how they march forward, each keeping its distance from that in front of it.

Now, if you have ever watched with care the ripples on a pond, you will know that though the ripples march forward, the water of which these ripples are composed does not—it merely rises up and down as each ripple comes by. The proof is simple. Throw in a bit of cork as a float. If the water were to flow along, it would take the cork with it. But no ; see how the cork rides the waves. It is the *motion* only that travels forward across the surface—the water simply swings to-and-fro, or rather up and down, in its place. Now that this has once been brought to your attention, you will be able to distinguish between the two kinds of movement—the apparent motion of the waves as they travel along the surface, and the actual motion of the particles in the waves, which is always of an oscillatory kind.

Here is a model of a wave-motion that will make the difference still clearer. At the top a row of little white

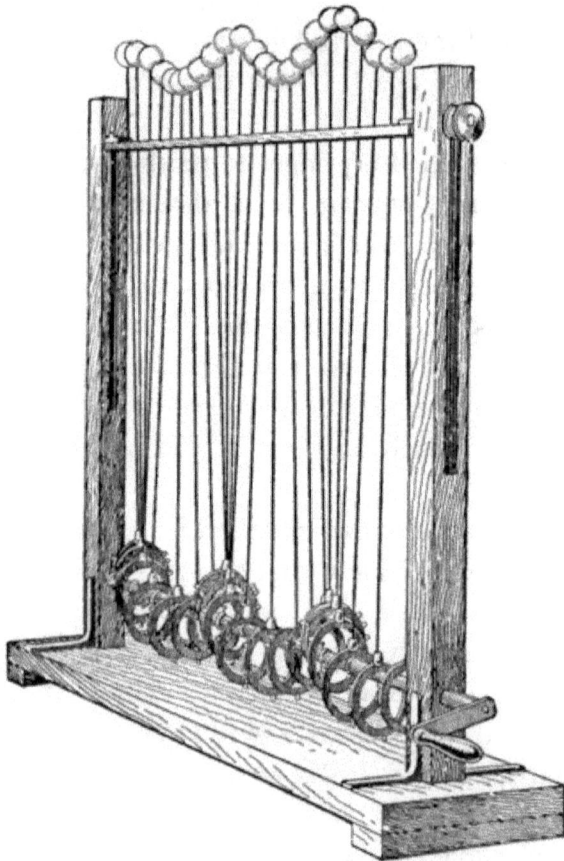

FIG. 2.

balls (Fig. 2) is arranged upon stems to which, in regular order one after the other, is given an oscillatory motion up and down. Not one of these white particles travels along. Each simply oscillates in its own place.

Yet the effect is that of a travelling wave, or rather set
of waves. The direction in which the wave travels is
transverse to the displacements of the particles. The
length from crest to crest of the waves is about 4 inches.
Their velocity of travelling depends, of course, on the
speed with which I turn the handle of the apparatus.
The amplitude of the displacement of each of the balls
is not more than one inch up or down from the centre
line.

Perhaps now you will be able to think of the little
wavelets of light, marching in ranks so close that there
are 40,000 or 50,000 of them to the inch, and having a
velocity of propagation of 185,000 miles a second.

Now let me state to you two important principles of
wave-motion—all-important in the right understanding
of the behaviour of waves of light.

(1) The first is that waves always[1] march at right
angles to their own front. This is how a rank of
soldiers march—straight forward in a direction square
to the line into which they have dressed. It was so
with the water-ripples that you have already seen.

(2) The second principle is that every point of any
wave-front may be regarded as a new source or centre
from which waves will start forward in circles. Look
at the sketch (Fig. 3). From P as a centre ripples
are travelling outward in circles, for there has been a
disturbance at P. Now if there is placed in the way of

---

[1] Always, that is to say, in free media, in gases, liquids, and
non-crystalline solids. In crystals, where the structure is such that
the elasticity differs in different directions, it is possible to have
waves marching obliquely to their own front.

these ripples a screen, S, or obstacle, with a hole in it,
all the wave-fronts that come that way will be stopped
or reflected back, except that bit of each wave-front
that comes to the gap in the screen. That particular
bit will go on into the space beyond, but will spread at
equal speed in all directions, giving rise to a new but
fainter set of ripples which will be again of circular form,

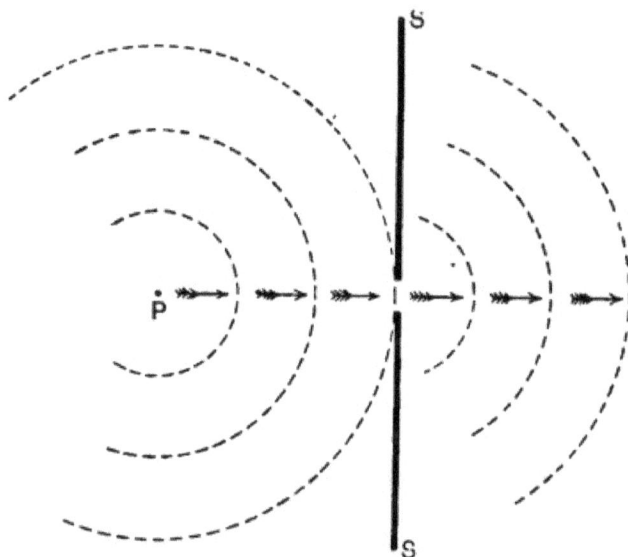

FIG. 3.

having their centre however not at P but at the gap in
the screen. This too I can readily illustrate to you in
my ripple-tank.

The first of these two principles is really a consequence of the second, and of another principle (that
of "interference") which concerns the overlapping of
waves. Of these we may now avail ourselves to find
how waves will march if we know at any moment the

shape of the wave-front. Suppose (Fig. 4) we knew
that at a certain moment the wave-front of a set of
ripples had got as far
as the curved line FF,
and that we wanted to
know where it would
be an instant later. If
we know how fast the
wave travels we can
think of the time taken
to travel some short
space such as half an
inch. Take then a pair
of compasses and open
them out to half an
inch. Then put the

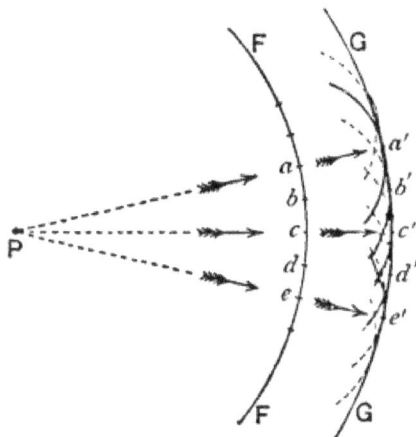

Fig. 4.

point of the compasses at some part—say *a*—of the curve
FF, and strike out the piece of circle as shown at *a'*.
That is where the disturbance would spread to in that
short interval of time if the bit of wave-front at *a* had
alone been allowed to spread forward. But the bit at *b*
is also spreading, so we must strike another arc, using *b*
as centre, and another at *c*, and another at *d*, and so on,
using the same radius for all of them. And now we see
that if all these bits, instead of acting each separately,
are acting at the same time, the wavelets from each will
overlap and give us one large enveloping curve at GG;
the effect being the same as though the wave-front FF
had itself marched forward to GG. Those parts of the
wavelets that tend to spread cross-ways in the overlapping balance one another; for instance, part of the

wavelet from *a* tends to cross downwards in front of *c*, while a part of the wavelet from *e* tends to cross upwards to an equal amount. These sideway effects cancel one another, with the result that the effect is the same on the whole as though the bit of wave at *c* had simply marched straight forward to *c'*.

Perhaps you will say that if this is true then when light-waves meet an obstacle some light ought to spread round into the shadow at the edges. And so it does as has already been said. But, owing to the exceeding smallness of the light-waves compared with the dimensions of ordinary objects, the spreading is so slight as to be unnoticed. In fact, except when we are dealing with the shadows of very thin objects, like hairs and pins, or with mere edges, the light behaves as though it simply travelled in straight lines.[1]

Our next business is to show how ripples can be made to diverge and converge. If we take a *point* as our source of the ripples, then they will of themselves spread or *diverge* from that point in all directions in circles, each portion of each wave-front having a bulging form. If we take as the source a flat surface, so as to get plane waves, they march forward as plane waves

---

[1] This is all that is meant by the old statement that light travels in "rays." There really are no rays. The harder one tries to isolate a "ray" by itself, by letting light go first through a narrow slit or pinhole, and then passing it through a second slit or pinhole, the more do we find it impossible ; for then we notice the tendencies to spread more than ever. If the word "ray" is to be retained at all in the science of optics, it must be understood to mean nothing more than the geometrical line along which a piece of the wave-front marches.

without either diverging or converging. If, however, we can in any way so manage our experiments as to get ripples with a hollow front instead of a bulging front, then the succeeding ripples will converge as they march. This is shown in Fig. 5. Suppose FF is a hollow wave-front marching forward toward the right. Think of the bit of wave-front at *a*. After a short interval of time it would spread (were it alone) to *a'*. Similarly *b* would spread to *b'*, and so on, so that when

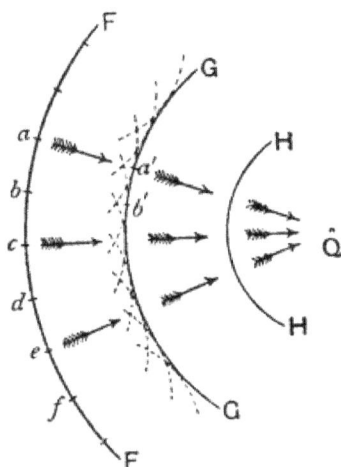

FIG. 5.

all these separate wavelets overlap, the effect is the same as though there the wave-front FF had marched to GG, closing in as it marches. After the lapse of another equally short interval it will have closed in to HH. It is clear that, on the principle that waves always march at right angles to their own front, they tend all to march inwards and meet at a new centre somewhere at Q. Suppose you ranged a row of soldiers in a curve like FF, and told each soldier to march straight forward between his comrades. If each soldier were to march at right angles to the curved line, they would all be marching toward a common centre, and would close in against one another!

Now it is obviously easy to make waves of light diverge—they do so of themselves if the source of light be a point. We shall see later how to make them converge; but, meantime, we will use what we know about divergence to help us to measure the relative brightness of two lights.

Here is a little electric glow-lamp. The shopman who sold it to me says that when it is supplied with electric current at the proper pressure,[1] it will give as much light as sixteen candles. I switch on the current and it shines. I light a standard candle,[2] so that you can compare the brightness for yourselves. Do you think that the glow-lamp is really sixteen times as bright as the candle? Your eye is really a very unreliable judge[3] of the relative brightness. We must, therefore, find some way of balancing the brighter and the less bright lights against one another. The instrument for doing this is called a *photometer*.

[1] Electric pressure, or "voltage," is measured in terms of the unit of electric pressure called the "volt." The usual electric pressure of the conductors which branch from the supply-mains into a house is 100 volts.

[2] The standard candle prescribed by the regulations of the Board of Trade as the legal standard of light in Great Britain is a sperm candle burning 120 grains of spermaceti per hour.

[3] This unreliability of the eye to form a numerically correct judgment is partly dependent on the physiological fact that the sensation is never numerically proportional to the stimulus. Though the stimulus be 16 times as great, the sensation perceived by the brain is not 16 times as great. The rule (Fechner's law) is that the sensation is proportional to the natural logarithm of the stimulus. The natural logarithm of 16 is 2·77; that is to say, the light that is 16 times as bright as 1 candle only produces a sensation 2·77 times as great. A single light of 100 candle brilliancy only produces a sensation 4·6 times as great as that of 1 candle. Besides this the iris diaphragm of the eye automatically reduces the size of the pupil when a brighter light shines into the eye, making the eye less sensitive.

But before we can understand the photometer we must first think about the degree of *illumination* which a light produces when it falls upon a white surface. I take here a piece of white cardboard one inch square. If I hold it close to my candle it catches a great deal of the light, and is brightly illuminated. If I hold it farther away it is less brightly illuminated. We can, therefore, alter the illumination of the surface by altering the distance. But we cannot use this principle for calculations about brightness until we know the rule that connects the distance with the degree of illumination ; and that rule depends upon the way in which light spreads when it starts from a point.

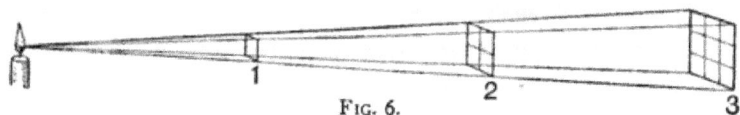

FIG. 6.

Suppose we think of the whole quantity of light that is spreading all round from a point. Of all that amount of light what fraction will be caught by this square inch of cardboard when I hold it a foot away? Not very much. But now think of that same amount of light as as it goes on spreading. Fig. 6 shows you that by the time that the light has travelled out from the centre to double the distance it will have spread (according to the law of rectilinear propagation discussed above) so that the diverging beam is now twice as broad each way. It will now cover a cardboard square that is 2 inches each way, or that has 4 square inches of surface. So if the same amount of light that formerly fell on 1 square inch is now spread over 4 square inches of surface, it follows that each of those 4 square inches is only illuminated one quarter as brightly as before. If you had a bit of butter to spread upon a piece of bread— and then you were told that you must spread the same piece of butter over a piece of bread of four times the surface, you know that the layer of butter would be only the quarter as thick! And so again, if I let the light spread still farther, by the time it has gone three times as far it will have spread over nine times the surface, and the degree of illumination on any one square inch at that treble distance will be only one-ninth part as great as at first. This is the so-called law of "inverse squares," and is simply the geometrical consequence [1] of the circumstance that the light is spreading from a point. Now we are ready to deal with the balancing of two lights. By letting two lights shine on a piece of card-board, or rather on two neighbouring pieces, and then altering the distance of one of the lights until both pieces of card are equally illuminated, we can get a balance of effects, and then calculate from the squares of the distances how bright the lights were. The eye, which is a very bad judge of relative unequal bright-nesses is really a very fair judge (and by practice can be trained to be a very accurate judge) of the equality of illumination of two neighbouring patches. But we must make our arrangements so that only one light shines

[1] The fact that a candle flame is not a mere point introduces a measurable error in photometry. It cannot be too clearly under-stood that the law of inverse-squares is never applicable strictly except to effects spreading from points. This criticism applies also to the use or misuse of the law of inverse-squares in magnetism and electricity.

upon each patch. One simple way of doing this is to let each light cast a shadow of a stick on a white surface, so that each light shines into the shadow cast by the other. If you alter the distances till the shadows are equally dark, then you know that the illumination of each is equal. But a better way is to arrange matters that the two illuminated patches are actually superposed. Here is a very simple and effective way of doing it. Two pieces of white cardboard, A and B (Fig. 7), forming a V-shape, are set upon a stand, between the two lights that are to be compared. One light shines upon the surface of A, and the other upon the surface of B. Through A are cut a number of slots or holes, so that the illuminated

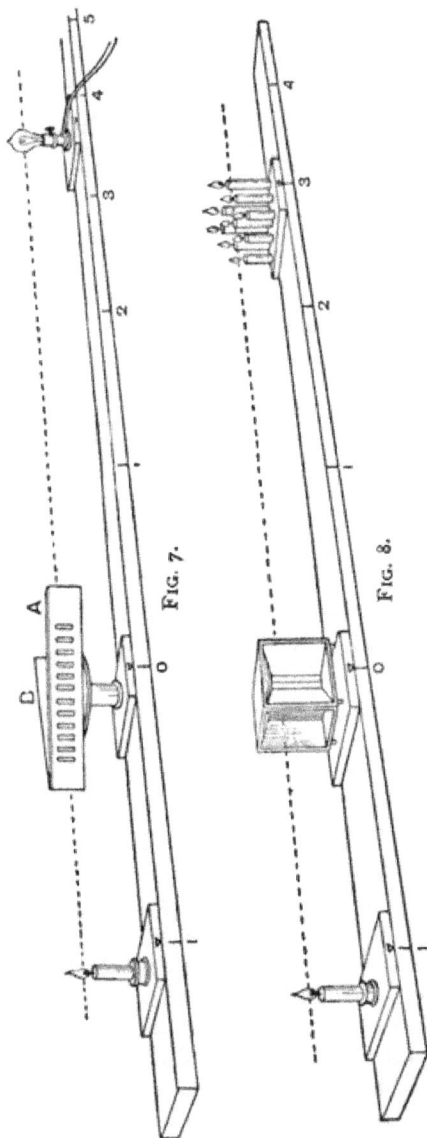

FIG. 7.

FIG. 8.

surface of B is seen through the slots in A. If the illumination of A is duller than that of B the slots will seem dark between the brighter bars of the front card; but if the illumination of A is brighter than that of B then the slots will seem bright between dull bars.[1] By moving one of the lights nearer or farther away, the respective illuminations can be altered until balance is obtained; and then the relative values are calculated from the squares of the distances. With this photometer let us now test our electric lamp to see if it is really worth sixteen candles. I put it on the photometer bench and move it backward and forward till the lights balance. You see it balances when rather less than four times as far away as the standard candle. It is, therefore, of not quite sixteen candle-power.

Another very simple and accurate photometer is made by taking two small slabs of paraffin wax (such as candles are made of) and putting them back to back

[1] This form of photometer is a modification by Mr. A. P. Trotter, M.A., of Cape Town, of the relief photometer invented in 1883 by the author and Mr. C. C. Starling. To prevent error arising from internal reflexion the back of the card A should be blackened. By setting the support at a fixed distance from the standard light on the left side, and altering, as needed to obtain balance, the distance of the light of which the brightness is to be measured, it is possible to make the instrument direct-reading; the scale to the right of the support being graduated so as to read not the actual distances but their squares. For instance, if the distance of the middle slot from the standard light be 1 metre, then on the other side the graduation must read 1 at 1 metre; 4 at 2 metres; 9 at 3 metres, and so forth. Accuracy of reading is promoted by the circumstance that when balance has been found for the middle slot of A the slots to the left of the middle will look darker, and those to the right brighter than the central one.

with a sheet of tin-foil or black paper between them.
They are then placed (as in Fig. 8) on the graduated
bench between the lights whose
brightness is to be compared together, and set in such a way that
one light shines on one paraffin
slab, and the other light on the
other slab, as in Fig. 9. If the
illuminations on the two sides
balance the edges of the slabs will seem equally bright.

FIG. 9.

But if the illumination on one face is stronger than
that on the other then that paraffin slab which is
more highly illuminated will seem brighter at its edge
than the other.[1] This is because of the translucent or

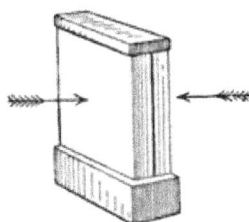

---

[1] This paraffin slab photometer is the invention of Dr. Joly,
F.R.S., of Dublin. It is an exceedingly satisfactory instrument.

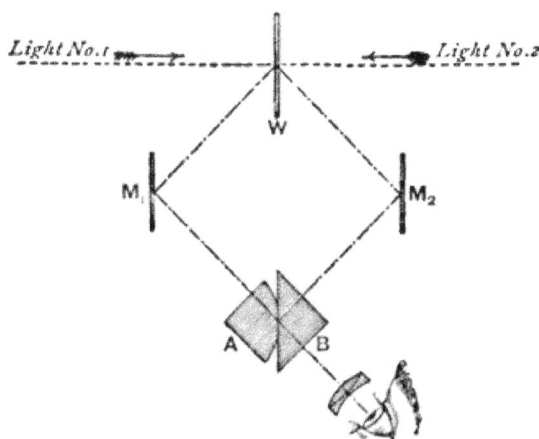

Light No.1   Light No.2

W

M₁   M₂

A   B

FIG. 10.

Either of these two forms of instrument here described is preferable
to the old-fashioned "grease-spot" photometer of Bunsen. But
both are surpassed in accuracy by the precision-photometer of

semi-opaque property of paraffin wax, which results in a diffusion of the light laterally. With this photometer it is very easy to balance the brightness of two lights, even if their tint be not quite identical. In Germany, they employ as standard, instead of a sperm candle, the little Hefner lamp filled with a chemical liquid known as amyl-acetate. But it has—as you see—the serious disadvantage of giving out a light which is unfortunately of a redder tint than most of our other lights. To be quite suitable, the lamp that we choose as a standard of light ought to be not only one that will give out a fixed quantity of light, but one that is irreproachable in the quality of its whiteness: it should be a standard of white light. Perhaps now that acetylene gas is so easily made it may serve as a standard, for as yet none of the proposed electric standards seem quite satisfactory.

Let us pass on to the operation of reflecting light by means of mirrors. A piece of polished metal such as

Brodhun and Lummer, which can, however, only be described here very briefly. It gives determinations that can be relied on to within one-half of one per cent. The two lights to be compared are caused to shine on the two opposite faces of a small opaque white screen, W (Fig. 10). The eye views these two sides, as reflected in two small mirrors, M_1 and M_2, by means of a special prism-combination, consisting, as shown, of two right-angled prisms of glass, A and B, which are cemented together with balsam over only a small part of their hypotenuse surfaces; the light from M_1 can pass direct through this central portion to the eye, but the uncemented portions of the hypotenuse surface of B act by total internal reflexion and bring the light from M_2 to the eye. The eye, therefore, virtually sees a patch of one surface of W surrounded by a patch of the other surface of W, and hence can judge very accurately as to whether they are equally illuminated or not.

silver, or a silvered glass, will reflect the waves of light, and so, though in an inferior degree, will any other material if only its surface be sufficiently smooth. By sufficiently smooth I mean that the ridges or scratches or roughnesses of its surface are decidedly smaller than the wave-length of the light. If the scratches or ridges on a surface are in width less than a quarter of the wave-length (in the case of light, therefore, less than about $\frac{1}{200000}$ inch) they do not cause any breaking up of

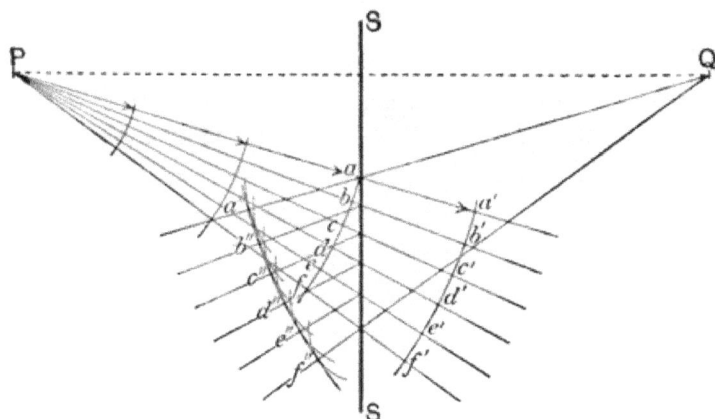

FIG. 11.

the waves; and such surfaces are, for optical purposes, quite "smooth." Indeed that is the usual way of polishing things. You scratch them all over with some sort of very fine powder that makes scratches finer than $\frac{1}{200000}$ of an inch.

Now the rebound of waves when they beat against a polished surface, whether that surface be a flat one or a curved one, can be studied by applying the same principles of wave-motion that we have already learned. In Fig. 11 we have light starting from a point at P and spreading. If a smooth obstacle, SS, is placed in the path of these waves they will meet it, but some parts of the wave-front will meet it before other parts. Think of the bit of the wave-front that meets the mirror at *a*. If it had not been stopped, it would after a brief moment of time have got as far as *a'*. But having bounded back from the surface it will set up a wavelet that will spread backwards at the same rate. Therefore, draw with your compasses the wavelet *a''*, using as radius the length *a a'*. The next bit of the wave-front *b* reaches the surface of the mirror a little later. The length from thence to *b'* is therefore a little shorter than *a a'*. So take that shorter length as radius and strike out the wavelet *b''*. Completing the set of wavelets in the same way we get the final curve of the reflected wave, which you see will now march backwards as though it had come from some point Q on the other side of the mirror. In fact, if the mirror is a flat one, Q will be exactly as far behind the surface as P is in front of it. We call the point Q the "image" of the point P. This reflexion of ripples as though they had come from a point behind the mirror I can show you by aid of my ripple-tank. I put in a flat strip of lead to serve as a reflector—see how the waves as they come up to it march off with their curvature reversed, as though they had started from some point behind the reflecting surface.

Again I can show you the same thing with a candle and a looking-glass. You know that we can test the direction in which light is coming by looking at the direction in which a shadow is cast by it. If I set up

(Fig. 12) this little dagger on a whitened board I can
see which way its shadow falls.   If now I place a candle
beside it on the board at P it casts a shadow of the
dagger on the side away from P.   Next, set up a piece
of silvered mirror glass a little farther along the board.
We have now two shadows.   One is the direct shadow
which was previously cast; the other is the shadow cast
by the waves that have been reflected in the mirror, and

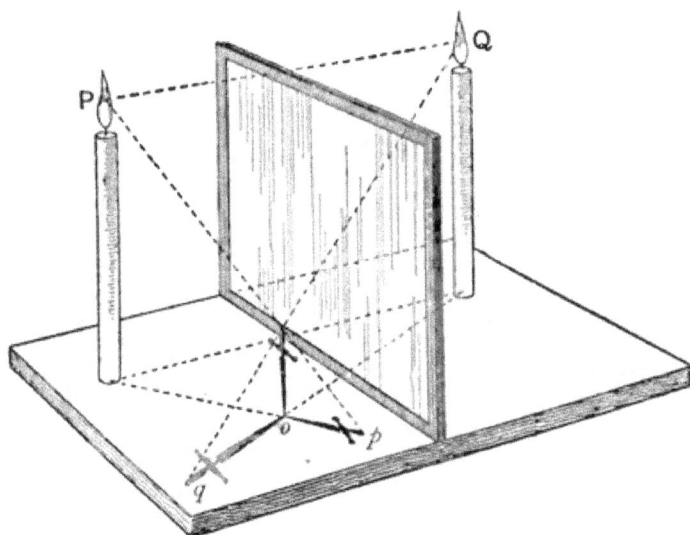

FIG. 12.

you see by the direction in which this second shadow
falls that it falls just as if the light had come from a
second candle placed at Q, just as far behind the mirror
as P is in front.   Let us put an actual second candle at
Q, and then take away the mirror, and you see the
second shadow in the same place and of the same shape
as before.   So we have proved by direct experiment that
our reasoning about the waves was correct.   Indeed,

you have only to look into a flat mirror, and examine
the images of things in it, to satisfy yourselves about
the rule. The images of objects are always exactly
opposite the objects, and are each as far behind the
mirror as the object is in front. Probably you have
all heard of the savage prince captured by sailors,
who, when he was taken on board ship and shown a
mirror hanging on a wall, wanted to run round to
see the other savage prince whom he saw on the other
side !

If instead of using flat mirrors we use curved ones,
we find different rules to be observed. That is because
the curved surfaces print new curvatures on the wave-
fronts, causing them to alter their lines of march. There
are, as you know, two sorts of curvatures. The surface
may bulge out—in which case we call it *convex ;* or it
may be hollowed—in which case we call it a *concave*
surface.

In my ripple tank I now place a curved piece of
metal with its bulging side toward the place where I
make the ripples. Suppose now I send a lot of plane
ripples to beat against this surface ; the part of the
wave-front that strikes first against the bulging curve is
the earliest·to be reflected back. The other parts strike
the surface later, and when reflected back have fallen
behind ; so that the ripples come back curved—the
curved mirror has, in fact, imprinted upon the ripples
a curvature twice as great as its own curvature. This
can be seen from Fig. 13, where we consider the straight
ripples marching to meet the bulging reflector. The
middle point M of the bulging surface meets the advan-

cing wave first and turns that bit back. If there had
been no obstacle the wave would, after a short interval
of time, have got as far as A. But where will it actually
go to? The bit that strikes M will go back as far as
B; the bit marked a will go on a little, and then be
reflected back. Take
your compasses again
and measure the dis-
tance it still has to go
to a', and then turn-
ing the compasses
strike out the arc a''.
Do the same for the
bits marked b and c,
and you will find the
overlapping wavelets

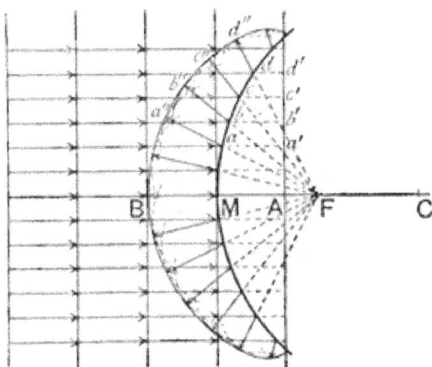

Fig. 13.

to give you the new outline of the reflected wave,
which marches backwards as though it had started from
the point marked F. This point F is half-way between
M and the centre of curvature of the surface. The
centre is marked C in the drawing.

So, again, if I use as reflector a hollow or concave-
curved surface, it will imprint upon the waves a concave
form, the imprinted curvature being twice as great as
the curvature of the reflecting surface. But now we
come upon a new effect. See in my ripple-tank how,
when the straight ripples beat against the concave
surface, so that the middle part of the wave-front is the
last to rebound, all the other parts have already re-
bounded and are marching back, the returning ripples
being curved inwards. In fact, you see that, being

themselves now curved ripples with hollow wave-fronts, they *converge* inwards upon one another, and march back toward the point F. A bit of the wave-front at P marches straight until it strikes the mirror at R. Then instead of going on to Q it is reflected inward and travels to F, toward which point other parts of the wave also travel. Here then we have found a *real* focus or meeting point of the waves; not, as in the preceding cases, a *virtual* focus from which the waves seemed to

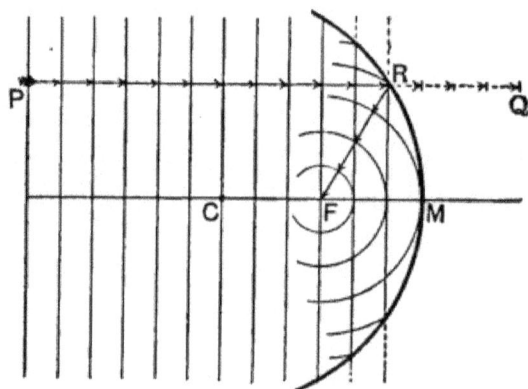

Fig. 14.

come. We have then learned that, for ripples at least, a concave mirror may produce a *real* convergence to a point.

Let us at once show that the same thing can be done with light-waves by using a concave silvered mirror.

From my optical lantern, with its internal electric lamp, my assistant causes a broad beam of light to stream forth. The air is dusty, and each little particle of dust catches a portion of the beam, and helps you to see which way it is marching. In this beam I hold a

concave silvered mirror. At once you see how by printing a curvature upon the waves it forces the beam to
converge (Fig. 15) upon a point here in mid-air. That
point is the focus. You will further notice that by
turning the mirror about I can shift the position of the

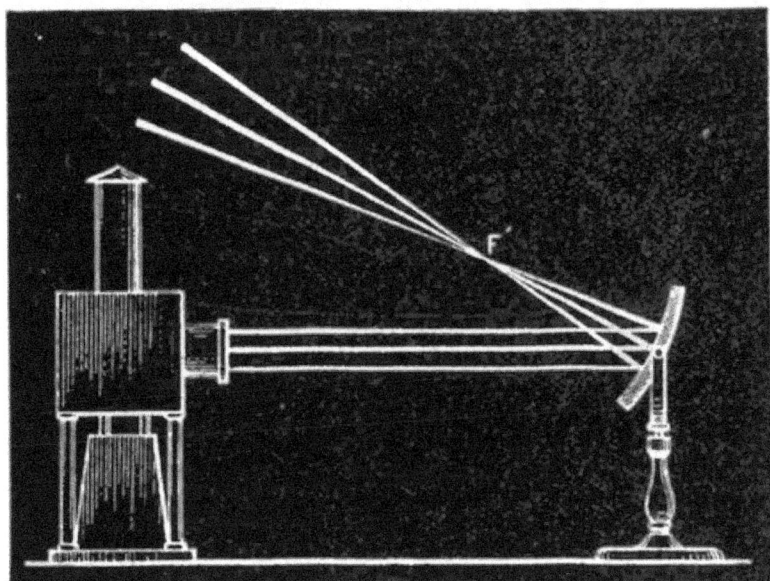

FIG. 15.

focus, and concentrate the waves in different places at
will.

If I replace the concave mirror by a convex one, I
shall cause a divergence of the waves. No longer is
there any real focus, but the waves now march away
as if they had come from a virtual focus behind the
mirror at F (Fig. 16), precisely as we saw for the ripples
in the ripple-tank.

We have now got as far as the making of real images

by so changing the shapes of the wave-fronts and their consequent lines of march as to cause them to converge to focal points. Let us try a few more experiments on the formation of images. Removing from the optical lantern all its lenses, let us simply leave inside it the electric lamp. You know that in this lamp there are two pencils of carbon, the tips of which do not quite touch, and

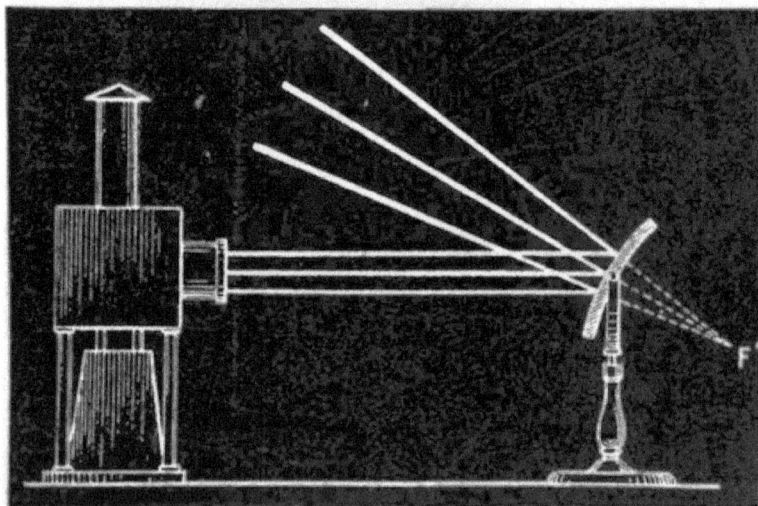

FIG. 16.

which are made white-hot by the flow of the electric current between them. I cover up the opening in front of the lantern by a piece of tin-foil, and in this I now stab a small round hole with a pointed stiletto. At once you see thrown on the screen an image (Fig. 17) of the two white-hot tips of the carbon pencils. The positive carbon has a flat end, the negative tip is pointed. That image is inverted as a matter of fact, and its formation on the screen is a mere consequence of the rectilinear propagation of the light. If I stab another hole we shall have another image. This time I have pierced a square hole, but the second image is just the same as the first, and does not depend on the shape of the hole. I pierce again a three-cornered hole—still another image. If I pierce a whole lot of holes I get just as many images, and they are arranged in a sort of pattern, which exactly corresponds to the pattern of holes I have pierced in the tin-foil.

Now if I wanted to produce one single bright image instead of a lot of little images scattered about, I must in

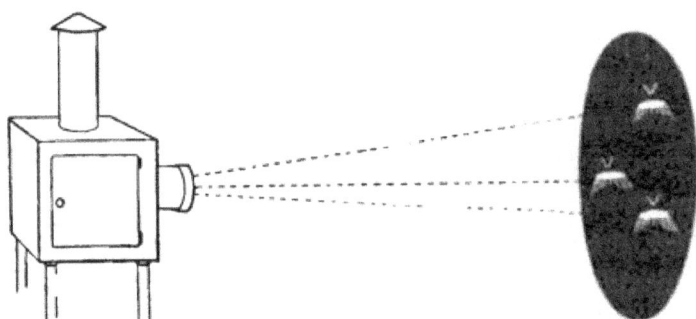

FIG. 17.

some way manage so to turn these various beams that they shall all converge upon the same region of the screen. In other words, the formation of bright images can be effected by using some appliance which will imprint a convergence upon the waves. You know that a concave mirror will do this. Very well, let me use a concave mirror. See how, when we choose one of the proper curvature to converge the light upon the screen, it blends all the images together, and gives us one bright image. We may remove our tin-foil cap altogether, so as to work with the whole beam, and we get a still more brilliant image of the carbon points.

Substituting for the arc-lamp a group of little coloured electric glow-lamps, I cause their beams to be reflected out into the room by my concave mirror, and here, by trying with a hand - screen of thin translucent paper, you see how I can find the real image of the group of lamps. This image is inverted ; and being in this case formed at a distance from the mirror greater than that of the object, it is magnified. If the object is removed to a greater distance the image comes still nearer in ; and is then of diminished size, though still inverted.

So far we have been dealing with the regular reflexion that takes place at properly polished surfaces. But if the surfaces are not properly polished—that is, if their ridges or scratches or roughnesses are not sensibly smaller than the size of waves, then, though they may still reflect, the reflexion is *irregular*. White paper reflects in this diffuse way. You do not get any definite images, because the slight roughnesses of the texture break up the wave-fronts and scatter them in all directions. That is why a white sheet of paper looks white from whichever aspect you regard it. If the substance is one which, like silk, has a definite fibre or grain that reflects a little better in one direction than in another, then the quantity of light reflected will depend partly upon the direction in which the grain catches the light, and partly upon the angle at which the light is inclined to the surface. This is easily demonstrated by examining the appearance of a piece of metal electrotyped in exact facsimile of a piece of silk fabric. Here is such a

piece. It was deposited[1] in a gutta-percha mould cast
upon a piece of figured silk brocade ; it reproduces the
exact shimmer of silk, because it reproduces the grain
of the silk in its operation of partial reflexion. If silk
is woven with warp of one colour and weft of another,
the different colours are better reflected at certain angles
—hence the effect produced by "shot" silk.

To illustrate the property of diffuse reflexion let me
show you two simple experiments. Here is a piece of
mirror. Upon it I paint with Chinese white the word
LIGHT. The letters look white on a dark background.
But if I use it to reflect upon the wall a patch of light
from the electric lamp the letters come out black. The
light that fell on those parts was scattered in all directions — so those parts looked white to you, but they
have diffused the waves instead of directing them
straight to the wall as the other smooth parts of the
surface do.

Let me prove to you how much light is really reflected
from a piece of paper. I have merely to shine my
lamp upon this piece of white paper, and hold it near
the cheek of this white marble bust for you to see for
yourselves what an amount of light it actually reflects
upon the object. Exchanging the white paper for a

[1] Made at the Technical College, Finsbury, by Mr. E. Rousseau,
instructor in electro-deposition. His process of casting, in a molten
compound of gutta-percha, the matrices, which are afterwards metal-
lised to receive the deposit in the electrotype bath, is distinctly superior
to the commercial process of taking moulds in a hydraulic press.
On one occasion he took for me a mould of a Rowland's diffraction
grating, having 14,400 parallel lines to the inch. Like the original
it showed most gorgeous diffraction colours.

piece of red paper,—that is to say of paper that reflects red waves better than waves of any other colour,—and you see how the red light is thrown back upon the bust, and brings an artificial blush to its cheek.

If light is reflected from one mirror to another one standing at an angle with the first, two or more images

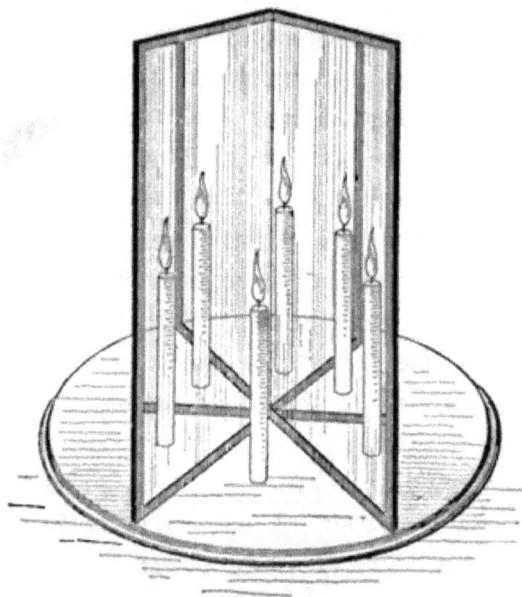

FIG. 18.

may be formed, according to the position of the mirrors. Here (Fig. 18) are two flat mirrors hinged together like the leaves of a book. If I open them out to an angle equal to one-third of a circle—namely, 120°—and then place a candle between them, each mirror will make an image, so that, when you peep in between the mirrors, there will seem to be three candles. If I fold the mirrors a little nearer, so that they enclose a quadrant of a circle

—or are at right angles—then there will seem to be four candles, one real one and three images. If I shut the angle up to 72°—or one-fifth of a circle—then there will seem to be five candles. Or to 60°—one-sixth of a circle—then there appear six candles. This is the principle of the toy called the *Kaleidoscope*, with which some most beautiful and curious combinations of patterns and colours can be obtained by the multiplication of images. Even with two such mirrors as these some quaint effects are possible. When nearly shut up, a single light between them seems to be drawn out into a whole ring of images. Open them out to 72° or to a right angle, and try the effect of putting your two hands suddenly between the mirrors. Ten hands or eight hands (according to the angle chosen) simultaneously appear as if by magic. Place between the mirrors a wedge of Christmas cake, and shut up the mirrors till they touch the sides of the wedge,—you will see a whole cake appear.

It is now time to pass on to another set of optical effects which depend upon the rate at which the waves travel. I have told you how fast they travel in the air—186,400 miles a second, or (if you will calculate it out by a reduction sum) one foot in about the thousand-millionth part of one second. Well, but light does not go quite so fast through water as through air—only about three-fourths as fast ; that is, it goes in water only at the rate of about 138,000 miles a second, or only about nine inches in the thousand-millionth part of a second. And in common glass it goes still slower. On the average—for glasses differ in their composition, and

therefore in the retardation they produce on light-waves
—light only goes about two-thirds as fast as in air. That
is, while light would travel one foot through air, it would
only travel about eight inches through glass.

Now as a consequence of this difference in speed
it follows quite simply that if the waves strike obliquely
against the surface of water or of glass that part of the
wave-front that enters first into the denser medium
goes more slowly, and the other part which is going on
for a little longer time though air gains on the part that
entered first, and so the direction of the wave-front is
changed, and the line of march is also changed. Let
us study it a little more precisely. If waves of light
proceeding from a point
P strike against the top
surface of a block of
glass, as in Fig. 19, how
will the retardation that
they experience on enter-
ing affect their march ?
Suppose that at a certain
moment a ripple has got
as far as FF′. If it had
been going on through
air it would, after a very

FIG. 19.

short interval of time, have got as far as GG′. But
it has struck against the glass, and the part that goes
in first instead of going as far as G′ will only get
two-thirds as far. So once more take your compasses,
and strike off a set of arcs for the various wavelets,
in each case taking as the arc two-thirds of the dis-

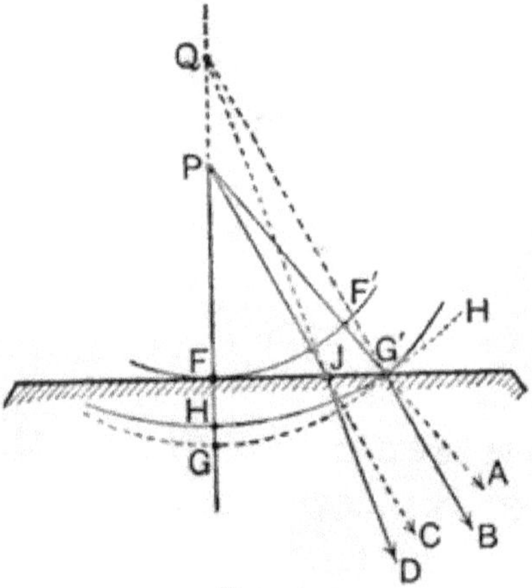

tance that the light would have had to go if after passing the surface it could have gone on to GG'. The overlapping wavelets build up the new wave-front HG', which you notice is a flatter curve, and has its centre somewhere farther back at Q. In fact, the effect of the glass in retarding the wave is to flatten its curvature and alter its march, so that in going on through the glass it will progress as though it had come not from P, but from Q, a point $1\frac{1}{2}$ times as far away. Consider the bit of wave-front that has been marching down the line PG'. When it enters the glass its line of march is changed—instead of going on along G'A it goes more steeply down G'B, as though it had come from Q. This abrupt change of direction along a broken path, caused by the entrance into a denser [1] medium, is known by the term *refraction*. Glass refracts more than water does; heavy crystal glass (containing lead) refracts more than the light sorts of glass used for window-panes and bottles; while many other substances have a still higher refractivity.

Now, we can use this property of the refracting substances to produce convergence and divergence of lightwaves, because, as you see, when we want to imprint a curvature on the wave-fronts, we can easily do this by using the retardation of water or of glass. Suppose we wanted to alter a plane-wave so as to make it converge to a focus, what we have got to do is to retard the middle part of the wave-front a little, so that the other

---

[1] "Denser," in its optical sense, means the same thing as *more retarding*. Compare with what is said on p. 62 in the Appendix to this Lecture.

parts shall gain on it. It will then be concave in shape,
and therefore will march to a focus. What sort of a
piece of glass will do this? A mere window-pane will
not. A thick slab will not, seeing it is equally thick all
over. Clearly it must be a piece of glass that is thicker
at one part than another. Well, suppose we take a
piece of glass that is thicker in the middle than at the
edges, what will it do? Suppose that, as in Fig. 20,
we have some plane-waves coming along, and that we
put in their path a piece of glass that is flat on one face

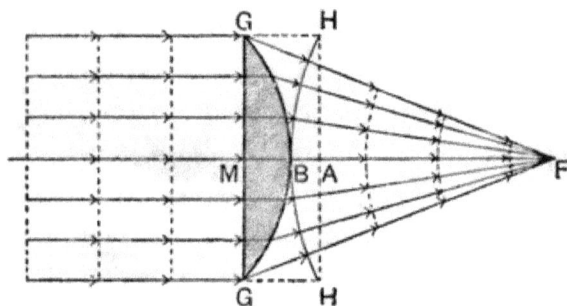

Fig. 20.

and bulging on the other face. Think of the time when
a wave-front has arrived at GG. A moment later where
will it be? The middle part that strikes at M will be going
through glass to B. This distance MB we know will be
only two-thirds as great as the distance to which it would
go in air. Had it gone on in air it would have gone as
far as A, the length MA being drawn $1\frac{1}{2}$ times as great
as MB. The edge parts of the wave-front go almost
wholly through air, and will gain on the middle part.
So the new wave-front, instead of being flat through
HAH, will be curved concavely in the shape HBH;

and as a result the wave will march on converging to
meet at F in a real focus.[1]   It would be the same if the
piece of glass were turned round the other way, with its
bulging face toward the light; it would still imprint a
concavity on the advancing wave and make it converge
to a focus.   This is exactly how a burning-glass acts.

With my ripple-tank I am able to imitate these
effects, but not very accurately, because the only way I
have of *slowing* the ripples is to make the water shal-
lower where retardation is to be produced.   This I do
by inserting a piece of plate glass cut to the proper shape.
Where the ripples pass over the edge of the submerged
piece of glass they travel more slowly.   Where they meet
the edge obliquely the direction of their march is changed
—they are *refracted*.   Where they pass over a lens-
shaped piece they are converged toward a focus.

It is, however, more convincing to show these things
with light-waves themselves.   Let me first show you
refraction upon the optical circle by the aid (Fig. 21) of
a special apparatus [2] for directing the beam toward the
centre at any desired angle.   Placing a large optical
circle with its face toward you and its back to the lantern,
I can throw the light obliquely upon the top surface of

---

[1] From Fig. 20 it is easy to see that the curvature of the impressed HAH is just half (if $MB = \frac{2}{3} MA$) of the curvature of the
glass surface.   Hence it follows that the focal length of the plano-
convex lens (if of glass having a refractivity of $1\frac{1}{2}$) is equal to twice
the radius of curvature of the lens-surface.   In the case of double-
convex lenses, each face imprints a curvature upon the wave as it
passes through.   See Appendix to Lecture I. p. 65.

[2] This apparatus, which can be fitted to any ordinary lantern,
consists of three mirrors at 45° carried upon an arm affixed to a

a piece of glass, the under surface of which has been
ground to a semi-cylinder
(Fig. 22). The refracted
beam emerges at a differ-
ent angle, its line of march
having been made more
steeply oblique by the
retardation of the glass.
If you measure the angles
not in degrees but by the
straight distances across
the circle, you will find
that, for the kind of glass
I am using, the proportion between the length CD (the

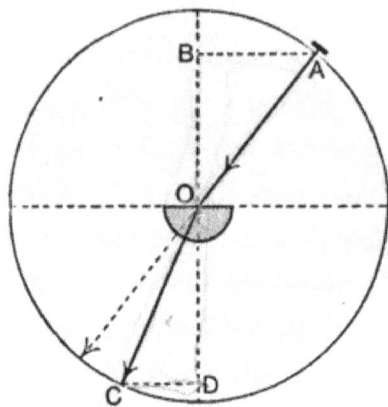

FIG. 22.

sleeve that fits the condenser-tube, as in Fig. 21. The beam after
three reflexions comes radially back across the axis of the con-

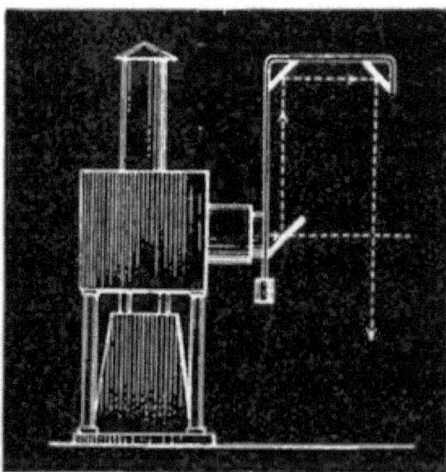

FIG. 21.

densers ; and by turning the arm around in the condenser-tube can
be used at any angle.

sine of refraction) and the length AB (the sine of incidence) is always just the proportion of 2 to 3, whatever
the obliquity of the incident beam.   When the incident
beam falls at grazing incidence most of it is reflected
and never enters the glass, and the part that does enter
is refracted down at an angle known as the critical or
limiting angle.

With this same optical circle I am able to show you
another   phenomenon,
that of *total internal re-*
flexion.   If I send the
light upwards through the
glass   hemisphere   (Fig.
23), at an angle beyond
that of the critical angle,
none of it will come up
through the surface ; all
will be reflected internally at the under side,
the top surface acting as

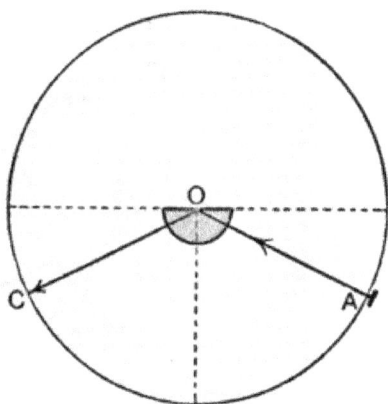

FIG. 23.

a polished mirror.   You can see the same effect with
a tumbler full of water with a spoon in it.

This same phenomenon of total reflexion can be
beautifully illustrated by the luminous cascade or fairy
fountain.   I allow water to stream out of a nozzle, and
shine light in behind through a window into the cistern
from which the water flows.   It falls in a parabolic curve,
the light following it internally down to the place where
the jet breaks (Fig. 24) into drops.

Total reflexion can also be illustrated by shining
light into one end of a solid glass rod, along which,

though it is of a bent and crooked shape, the light travels until it comes to the other end.

Returning now to the use of lenses to cause the waves to converge and diverge, we will adjust our lantern to send out a straight beam, and then interpose in

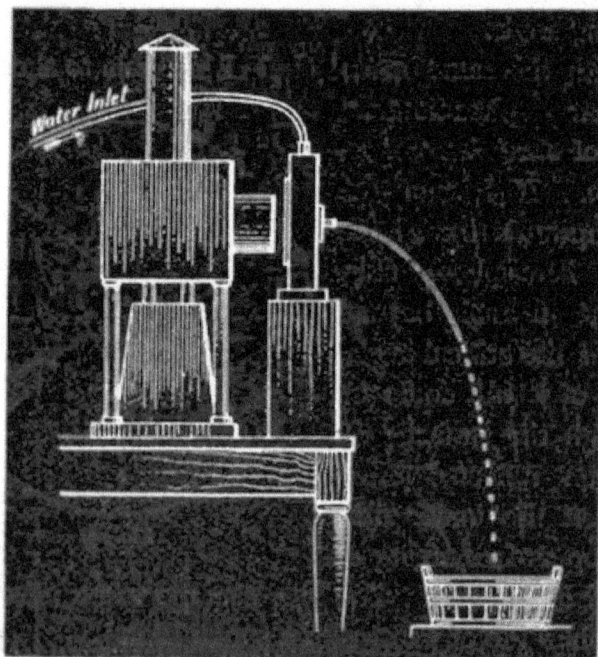

FIG. 24.

the path a lens made of glass thicker in the middle than at the edges. At once it is observed—thanks to the dust in the air—to make these waves converge to a focus at F (Fig. 25). This is again a real focus. A lens that is thus thicker in the middle than at the edges is called a *convex lens*.

Had we taken a piece of glass that is thinner in the

middle than at the edges—a *concave* lens—the effect

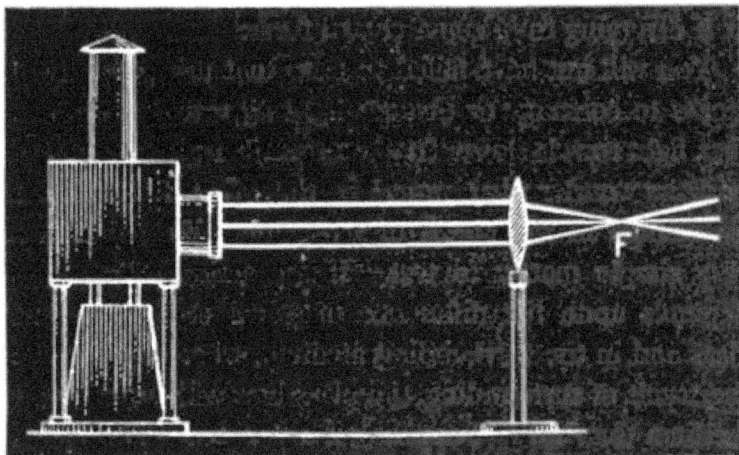

FIG. 25

would be the opposite.   Since the thin middle retards
the mid parts of the wave-front less than the thick glass

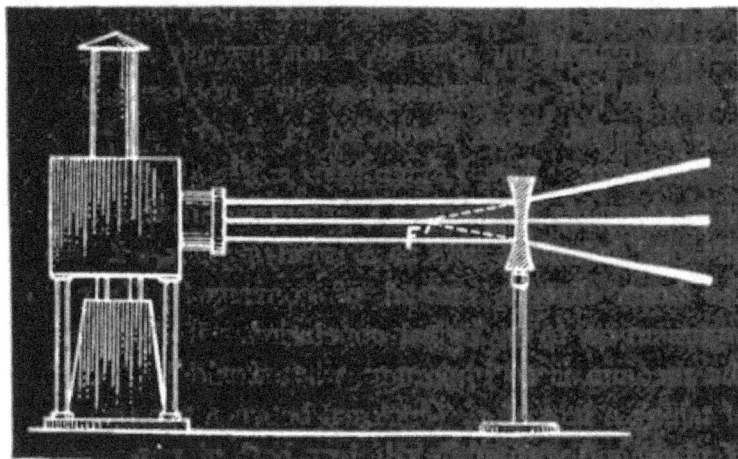

FIG. 26.

edges retard the edge parts, the middle part of the

wave-front will gain on the outlying parts, and the wave will emerge as a bulging wave, and will therefore march as if diverging from some virtual focus.

You will not have failed to note that this property of lenses to converge or diverge light depends on the fact that light travels slower in glass than in air; and you will perhaps wonder what would be the effect if there were no change in the speed of travelling. Well, that is a very simple matter to test. If the action of the lens depends upon the difference of speed of light in the glass and in the surrounding medium, what ought to be the result of surrounding the glass lens with some other medium than air? Suppose we try water. The speed of light in water is less than in air—it is more nearly like that in glass. And if the action depends on difference of speed, then a glass lens immersed in water ought to have a less action than the same glass lens in air. Try it, and you see at once that when immersed in water a magnifying glass does not magnify as much as it does in air. A burning-glass does not converge the rays so much when immersed in water; its focus is farther away. Nay, I have here a lens which you see unquestionably magnifies. I immerse it in this bath of oil—and behold it acts as a minifying lens—it makes the beam diverge instead of converge! Carry the argument on to its logical conclusion. If the effect of the medium is so important, what would be the effect of taking a lens of *air* (enclosed between two thin walls of glass) and surrounding it by a bath of water or oil? If the reasoning is right, a concave air lens in oil ought to act like a convex glass lens in air, and a convex air

lens in oil like a concave glass lens in air. Let us put
it to the test of experiment. Here is a concave air lens.
In air it neither converges nor diverges the light—the
speed inside and outside the
lens is the same—therefore
there is no action. But plunge
it in oil (Fig. 27) and, see, it
brings the beam to a focus
(F) exactly as a convex glass
lens in air would do.

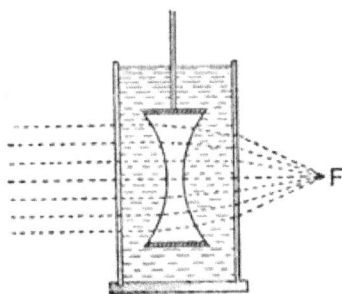

FIG. 27.

Let me sum up this part
of my subject by simply
saying that lenses and curved mirrors can change the
march of light-waves by imprinting new curvatures on
the wave-fronts. Indeed, speaking strictly, that is all
that any lens or mirror, or combination of lenses or of
mirrors, can do.

Now the human eye, that most wonderful of all
optical instruments, is a combination of lenses within a
cartilaginous ball, the back of which is covered on its
inner face with an exquisitely fine structure of sensitive
cells, through which are distributed ramifications of the
optic nerve. All that that nerve can do is to feel the
impressions that fall upon it and convey those impressions to the brain. All else must be done on the one
hand by the lens-apparatus that focuses the waves of light
on the retina, or on the other hand by the brain that is
conscious of the impressions conveyed to it. With neither
the nerve-structures nor with the brain are these lectures
concerned. We have merely to treat of the eye as a
combination of lenses that focuses images on the retina.

Consider a diagram (Fig. 28) of the structures of the human eyeball. The greater part of the refractive effect. is accomplished by a beautiful piece of transparent horny substance known as the *crystalline lens* (L_2), which is situated just behind the *iris* or coloured diaphragm of the eye. The pupil, or hole through the iris, leads straight toward the middle of this crystalline

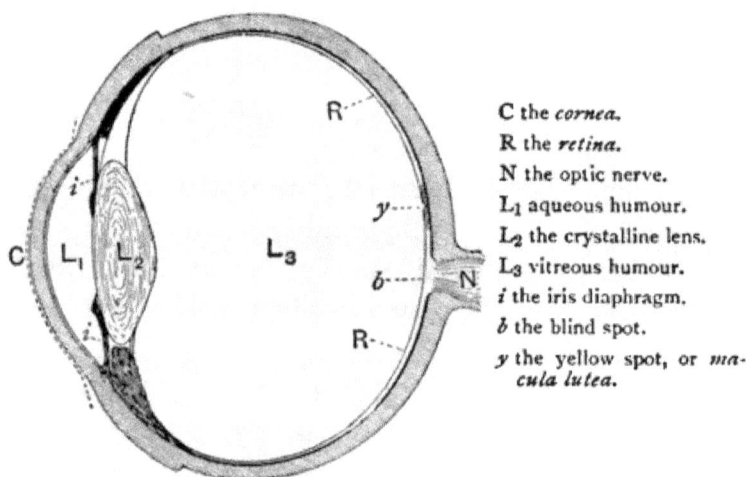

C the *cornea*.
R the *retina*.
N the optic nerve.
L_1 aqueous humour.
L_2 the crystalline lens.
L_3 vitreous humour.
i the iris diaphragm.
b the blind spot.
y the yellow spot, or *macula lutea*.

FIG. 28.

lens. But it is immersed in a medium, or rather between two media, a watery medium (L_1) in front and a gelatinous one (L_3) behind; the latter filling up the rest of the globe of the eyeball. The crystalline lens has therefore a less magnifying power than it would have if it were immersed in air. It acts very much as a lens in water. But the watery liquid in front of it also acts as a lens, since it occupies the space in front of the crystalline lens and between it and the transparent *cornea*, the bulging window of the eye. Taken together these form a lens-combination adapted to form images upon that back-screen or *retina*, R, where are spread out the sensitive nerve structures. All that the eye can do as an optical instrument can be imitated by optical combinations of lenses. An ordinary photographic camera may be regarded as a sort of artificial eye. In front is a combination of lenses the function of which is to focus images upon a back screen, or upon a plate which is made chemically sensitive. To make the analogy more complete one ought to think of the eye as a kind of camera in which the hollow body is filled up with a thin transparent watery jelly, and in which also the space between the front lens and the one behind it is full of water.

Apart from the complication introduced by the watery and gelatinous media, it is very easy to imitate the optical arrangements of the eye by lenses. Any photographic camera will serve indeed for the purpose. Its lens combination throws upon the screen at the back real images of the objects placed in front.

As in the camera, so in the eyeball, the images thrown on the back are inverted images. If you have not thought of this before it seems hard to believe it : nevertheless it is true. You have all your lives had the images inverted. Your brains, while you were yet babies learned to associate the impression received on the lower part of the retina with objects high above you. However you may explain or doubt, the facts are simply what they are : the images are upside-down at the back of your eyeball.

Beside the general proof afforded by camera-images, there are two extremely simple proofs of this fact. The first any of you can try at home; all the apparatus it needs being a common pin and a bit of card. It depends upon the circumstance that if you hold a small object close to a lens a shadow of it may be cast right through the lens without being turned upside down. Here is a lens—it will form inverted images of objects if it focuses them on a screen. But hold a small object close to the lens (Fig. 29) and shine light through it; the shadows are actually cast right side up on the screen.

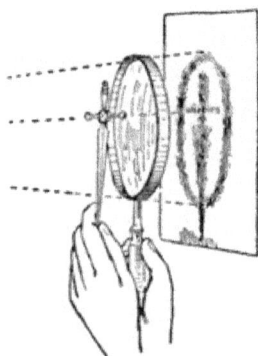

FIG. 29.

Now take a visiting-card and prick a pinhole through it with a large-sized pin. Place this hole about an inch from the eye and look through it at a white cloud or a white surface strongly illuminated. Then hold the pin upright, as in Fig. 30, between the eye and the pinhole. It may require a little patience to see it, as the pin must be held exactly in the right place. You know you are holding it with the head up, yet you see it with its head down, looking as in Fig. 31. Now if in the case where you know that its shadow is being thrown upright on the back of your eye you feel the shadow upside down, it follows that when you feel any image right way up it must really be an inverted image that you are feeling.

The other proof has the merit of being direct and objective, but does not succeed with every eye—some persons have the cartilaginous walls of the eyeballs too

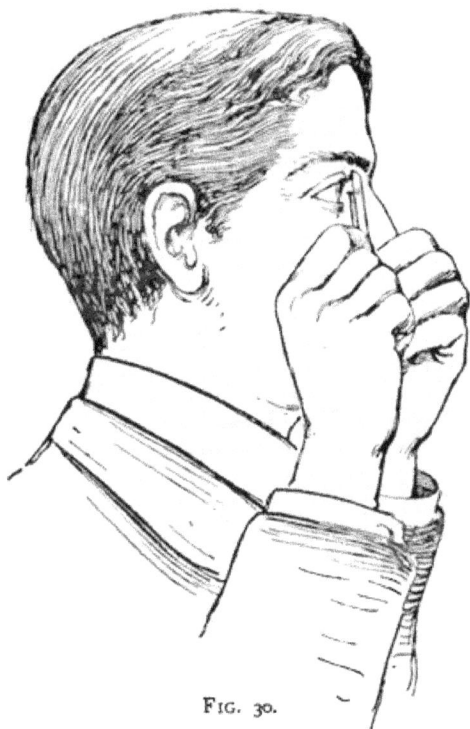

FIG. 30.

thick.   Stand in front of a mirror, close one eye—
say the right—and hold a candle in the hand on the
same side.   Hold the candle about
at the level of the closed eye so that
its light just falls across the bridge
of the nose into the open eye.
Then if you look very carefully you
will see, right in the extreme corner
of the eye, shining dimly through
the cartilaginous white wall, a small

FIG. 31.

image of the candle flame—*and it is inverted*.   If you

raise the candle higher, the image goes down; if you lower the candle, the image rises.

Leaving lenses let me show you a couple of interesting experiments depending on the property of refraction that we have been discussing. In passing through the earth's atmosphere obliquely, as they do when the sun is low down near the horizon, the sun's rays are refracted, and he seems to be a little higher up in the sky than he really is. Indeed, under certain circumstances, the sun can be seen above the horizon at a time when it is absolutely certain that he has really set; his rays in that case come in a curved path over the intervening portion of the globe. Now the circumstances in which this can occur are these—that the successive strata of the air must be of different densities; the densest below, next the earth, and the less dense above. To demonstrate this I will take a glass tank into which there have been carefully poured a number of solutions of chloride of calcium in water of

Fig. 32.

different densities — the densest at the bottom. You note that the beam of light sent into the trough takes a curved path (Fig. 32). In fact, the light turns round a corner.

The difference of refractivity that accompanies difference of density is well shown by a very simple experiment upon heated air. You all know that when air is heated it rises, becoming less dense. You all know that, when cooled, air becomes more dense, and tends to fall. But did you ever *see* the hot air rising

from your hand, or even from a hot poker? Or did
you ever *see* the cold air descending below a lump of
ice? This is exceedingly easy to show you. All I
require is a very small luminous point. We will take
the light of an arc-lamp, shining through a small hole
in a metal diaphragm close to it, and let it shine on the
white wall. Now I let this hot poker cast its shadow
on the screen, and you see torrents of hot air, which
rising, cast their shadows also. Here is a lump of ice.
The cold air streaming down from it casts its shadow.
Even from my hand you see the hot air rising. A
candle flame casts quite a dense shadow, and when I
open a bottle of ether you see the ether vapour—which
is ordinarily quite invisible—streaming out of the neck
and falling down. Even a jet of escaping gas reveals
itself when examined by this method.

Another curious experiment consists in using as a
lens a piece of glass which has been ground so as to be
curved only one way—say right and left—but not
curved in the other way. If this
lens is thicker in the middle part
from top to bottom, as in Fig. 33,
than it is at the two edges, it will
magnify things from right to left,
but not from top to bottom; hence
it produces a distortion. I throw

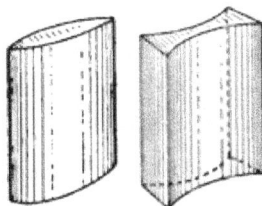

FIG. 33.

upon the screen the portrait of a well-known old gentle-
man. Then if I interpose in front of him one of these
"cylindrical" lenses, his face will be distorted. And if I
then turn the lens round the distortion will alternately
elongate his features and broaden them. There are

also cylindrical lenses of another kind, thinner in the middle than at the edges. These produce a distortion by minifying.

Finally, I return to the point which I endeavoured to explain to you a few minutes ago, that all that any lens or mirror can do is to impress a curvature upon the wave-fronts of the waves.

The most striking proof of this is afforded by that now rare curiosity the magic mirror of Japan. In old Japan, before it was invaded and degraded by Western customs, many things were different from what they now are. The Japs never sat on chairs—there were none to sit upon. They had no looking-glasses—their mirrors were all of polished bronze; and, indeed, those interesting folk had carried the art of bronze-casting and of mirror polishing to a pitch never reached in any other nation before them. The young ladies in Japan when they were going to do up their hair used to squat down on a beautiful mat before a lovely mirror standing on an elegant lacquered frame. Fig. 34 is photographed from a fine Japanese drawing in my possession. You may have seen pretty little Yum-yum in the "Mikado" squat down exactly so before her toilet-table. Here (Fig. 35) is one of these beautiful Japanese mirrors, round, heavy, and furnished with a metal handle. One face has been polished with care and hard labour; the other has upon it in relief the ornament cast in the mould—in this case the crest of the imperial family, the *kiri* leaf (the leaf of the *Paullonia imperialis*) with the flower-buds appearing over it. The polished face is very slightly convex; but on looking into it none of you young

FIG. 34 — Japanese Girls with Mirrors.

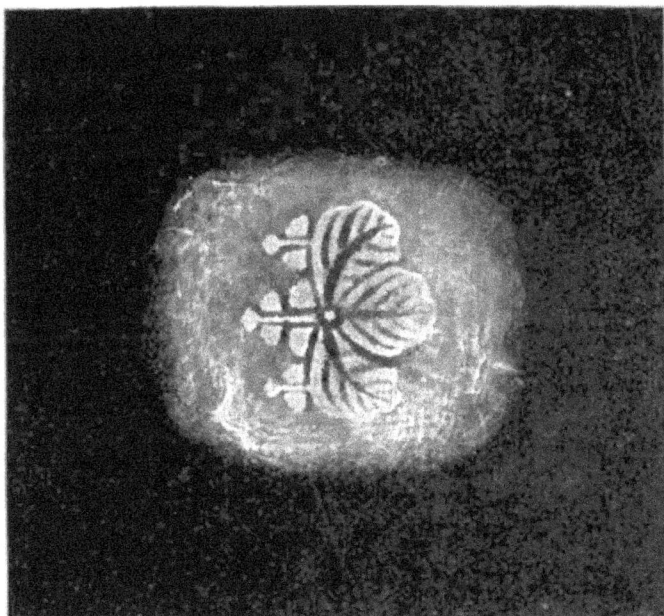

Fig. 36

Image reflected upon the wall by the polished front face.

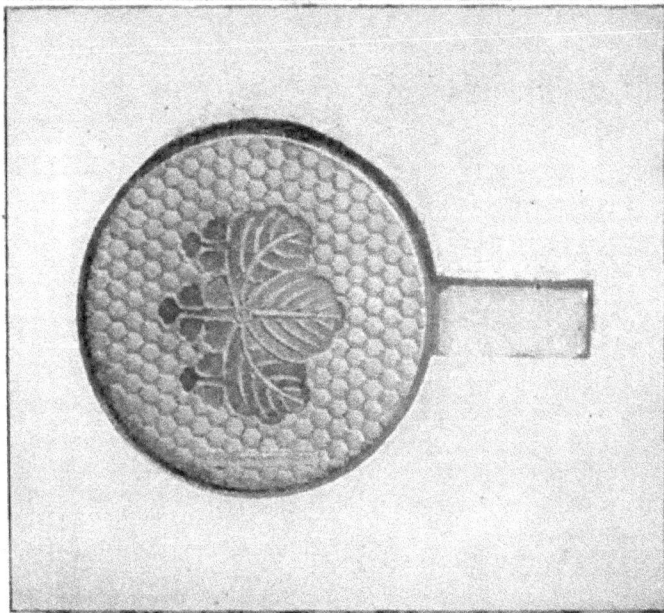

Fig. 35.

Japanese Mirror; showing the pattern cast in relief on the back.

ladies would see anything but your own fair faces, or the faces of your friends around you, or the things in the room. Certainly you would see nothing of the ornament on the back. It is merely—so far as you or the former owner of the mirror is concerned—a mirror.

But now take this mirror and hold it in the light of the sun, or in the beams of an electric lamp, and let it reflect a patch of light upon the white wall, or upon a screen. What do you see? Why, in the patch of light reflected from the front of the mirror, you see (Fig. 36) the pattern that is on the back. This is the extraordinary "magic" property that has made these mirrors so celebrated.

Another mirror has at the back a circle in high relief, with a fiery dragon in low relief sprawling around it. The face is beautifully polished, and shows no trace of the pattern at the back. But when placed in the beams of the arc-lamp it throws a patch of light on the wall, in which the circle stands out as a brilliant line, whilst the dragon is invisible. It is quite usual for the parts in high relief to produce this "magical" effect, while those in low relief produce none.

For many years it was supposed that these mirrors were produced by some trick. But the extraordinary fact was discovered by Professor Ayrton in Japan that the Japanese themselves were unaware of the magic property of the mirrors. It results, in fact, from an accident of manufacture. Not all Japanese mirrors show the property: those that show it best are generally thin, and with a slightly convex face. It was demonstrated by Professor Ayrton, and I have since accumu-

lated some other proofs,[1] that the effect is due to
extremely slight inequalities of curvature of surface.
These arise accidentally in the process of polishing.
The mirrors are cast in moulds.   To polish their faces

FIG. 37.

they are laid down on their backs by the workman, who
scrapes them violently with a blunt iron tool, using great
force.   Fig. 37 is taken from a Japanese print in the
British Museum.   During this process they become
slightly convex.   The polishing is completed by scouring

---

[1] These differences of curvature of surface can be proved (1) by
actual measurement, in some cases by spherometer , (2) by placing
a convex lens in front to correct the general convexity and then
observing directly, as in Foucault's method for testing figure of
mirrors ; (3) by reflecting in the mirror the image of a number of
fine parallel lines, whose distortion reveals the inequalities of curvature ; (4) by taking a mould in gutta-percha, and reproducing the
polished surface by electrotype, which is then silvered.   The silvered
type will act as a magic mirror.   In some cases the "silvering"
wears off the surface unequally, remaining last on the parts that are
slightly concave.   The front then shows faintly to the eye the
pattern on the back.

with charcoal and scrubbing with paper, after which they are "silvered" by application of an amalgam of tin and mercury. Now during the violent scraping with the iron tool the mirror bends, but the thin parts yield more under the pressure than the thick parts do; hence the thick parts get worn away rather more than the thin parts, and remain relatively concave, or at least less convex.

Amongst the proofs that these very slight inequalities of curvature can thus reveal themselves by imprinting a convergivity or a divergivity upon the reflected waves, let me show you this glass mirror, silvered in front and quite flat, but having a star engraved on its back. By merely blowing air against the back to bend it, the star becomes visible in the patch of light reflected from the face. Here the thin parts yield more than the thick ones. Again, simply heating a piece of looking-glass locally, by applying a heated iron stamp to the back of it, will cause the glass to expand in the heated region, and exhibit the pattern of the stamp in the patch of light reflected on the wall by the mirror.

Lastly, I have to exhibit some magic mirrors made by a former pupil of mine, Mr. Kearton—English magic mirrors—which have no pattern upon them, either back or front, but yet show images in the light they reflect upon the wall. Here is one that shows a serpent; here another with a spider in his web; another with a man blowing a horn. These are made by etching very slightly upon the brass mirror with acid (an immersion of three seconds only is ample), and then polishing away the etched pattern. After polishing for twenty minutes the pattern will have disappeared entirely from sight. But you may go on polishing for an hour more, and still the minute differences of curvature that remain will suffice,—though quite undiscoverable otherwise—to produce a magic image in the patch of reflected light. Though so excessively minute these differences of curvature of the mirror print their form upon the wave-fronts of the light, and alter the degree of convergency or divergency of the beam.

APPENDIX TO LECTURE I

General Method of Geometrical Optics

THE method of teaching Geometrical Optics upon the lines of the wave-theory, which is the key-note to this Lecture, has been followed systematically by the author for fifteen years in his regular courses of instruction in Optics to students attending his lectures in Physics. The treatment of the subject before the audience attending the Christmas course at the Royal Institution, many of whom were juveniles, was necessarily simplified and popularised ; but the essential features of the method remain.

The author also published a brief notice of this method of teaching the subject in 1889 in a paper entitled "Notes on Geometrical Optics," read before the Physical Society of London, and printed in the *Philosophical Magazine* (October 1889, p. 232).

As the development of the method in the present lecture is so slight, the author deems it expedient to add as an Appendix a few further points showing the application to the establishment of formulæ for lenses and mirrors. These are, in fact, established much more readily on this basis than by the cumbrous methods that are consecrated by their adoption in every text-book of Geometrical Optics.

Basis of the Method

In treating optics from the new standpoint, we have to think about surfaces instead of thinking about

mere lines. Waves march always at right angles to
their surfaces; a change in the form of the surface
alters the direction of march. The wave-surface is
to be considered instead of the "ray." The curvature
of the surface therefore becomes the all-important con-
sideration. *All that any lens or mirror or any system
of lenses or mirrors can do to a wave of light is to imprint a curvature upon the surface of the wave.* If the
wave is initially a plane-wave, then the curvature imprinted
upon it by the lens or mirror will result in making it either
march toward a point (a real focus) or march as from a
point (a virtual focus). If the wave possesses an initial
curvature, then all that the lens or mirror can do is to
imprint another curvature upon its surface, the resultant
curvature being simply the algebraic sum of the initial and
the impressed curvatures. As will be seen, in the new
method the essential thing to know about a lens or mirror
is the curvature which it can imprint on a plane wave:
this is, indeed, nothing else than what the opticians
call its "power"; the focal power being inversely proportional to the so-called focal length. Another but less vital
point in the method, is the advantage of using instead of
the so-called index of refraction a quantity reciprocally related to it, and here denominated the velocity-constant.
The use of the index of refraction dates from a time
anterior to the discovery that refraction was a mere
consequence of the difference of velocity of the
waves in different media. The index of refraction
is a mere ratio between the sines (or originally the
cosecants) of the observed angles of incidence and refraction. The uselessness of clinging to it as a founda-
tion for lens formulæ is shown by the simple fact that, in
order to accomplish the very first stage of reasoning in the
orthodox way of establishing the formulæ, we abandon the
sines and write simply the corresponding angles, as Kepler
did before the law of Snell was discovered. The ele-
mentary formulæ of lenses are, in fact, where Kepler left
them. It is now common knowledge that the speed of
light, on which refraction depends, is less in optically dense

media than in air. The speed of light in air is not
materially different from one thousand million feet per
second, or thirty thousand million centimetres per second.
If we take the speed of light in air as unity, then
the numeric expressing the speed in denser media, such as
glass or water, will be a quantity less than unity, and will
differ for light of different wave-lengths. It is here preferred to take the speed of light in air, rather than *in vacuo*,
as unity, because lenses and optical instruments in general
are used in the air. The numeric expressing the relative
velocity in any medium is called its " velocity-constant " ;
it is the reciprocal of the index of refraction. The velocity-
constant, for mean (yellow) light, for water is about 0·75 ;
that of crown glass 0·65 ; that of flint glass from 0·61
to 0·56, according to its density.

Method of Reckoning Curvature

The Newtonian definition of curvature as the reciprocal of
the radius has a special significance in the present method of
treating optics : for some of the most important of lens and
mirror formulæ consist simply of terms which are reciprocals
of lengths, that is to say of terms which are curvatures. The
more modern definition of curvature as rate of change of
angle per unit length of the curve (Thomson and Tait's
Natural Philosophy, vol. i. p. 5) is equivalent to Newton's ;
for if in going along an arc of length δs, the direction
changes by an amount $\delta \theta$, the curvature is $\delta \theta / \delta s$. But the
angle $\delta \theta = \delta s / r$, where r is the radius of curvature ; hence
the curvature $= \delta s / r \delta s = 1 / r$.

There is, however, another way of measuring curvature,
which, though correct only as a first approximation, is
eminently useful in considering optical problems. This
way consists in measuring the bulge of the arc subtended
by a chord of given length.

Consider a circular arc AP, having O as its centre.
Across this arc draw a chord PP' of any desired length.
The diameter AB bisects it at right angles in M. The

short line MA measures the depth of the curve from arc
to chord. If the radius is taken as unity the line MA is the
versed-sine of the angle subtended at B by the whole chord,
or is the versed-sine of the semi-angle subtended at the
centre. In Continental works
it is frequent to use the name
sagitta for the length of this line
MA ; and as this term is preferable to versed-sine, and can
be used generally irrespective of
the size of radius, it is here
adopted. The proposition is
that, for a given chord, the
sagitta is (to a first degree of approximation) proportional
to the curvature. For it follows from the construction
that

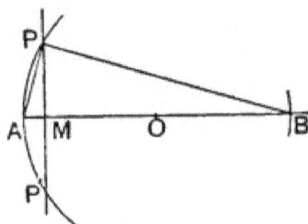

FIG. 38.

$$MA . MB = (PM)^2 ;$$

assuming PM as unity,

$$MA = \frac{1}{MB} = \frac{1}{2r - AM} .$$

But, for small apertures, AM is small compared with $2r$,
and may be neglected in the denominator, whence, to a
first approximation,

$$MA = \frac{1}{2} . \frac{1}{r} .$$

Twice [1] the sagitta represents numerically the curvature.
The error is less than one per cent when the semi-angle subtended at the centre is $10°$; less than two per cent when
it is $15°$; less than five per cent when it is $25°$.

If the method of reckoning curvatures by means of the
sagitta required justification, that is afforded by the fact
that the practical method of measuring the curvatures of

---

[1] Though the sagitta is numerically *half* the curvature, since all
the formulæ of first approximation are homogeneous and of the first
degree as regards sagittæ and curvatures, the numerical factor ½ dis-
appears in passing from sagittæ to curvatures, or *vice versâ*.

lenses and mirrors by the *spherometer* consists essentially in applying a micrometer-screw to measure the sagitta of the arc subtended by a fixed chord, the diameter of the contact circle drawn through the three feet of the instrument.   In this case, as indeed in all cases where accuracy, not approximation, is desired, the basis for calculation of the correction exists in the actual size of the diameter of the contact circle, which is a fixed parameter for all measurements made with the instrument.   The "*lens measurer*" used by opticians to test the curvatures of spectacle-lenses is a very simple micrometer which reads off directly the sagitta of the curve against which it is pressed, and indicates on a dial the value in terms of formula [10] on p. 65.

The sign of the curvature remains to be defined.   In the case of actual waves of light, the sign adopted will be + for the curvature of waves which are converging upon a real focus ; – for those which are diverging either from a luminous source or from a virtual focus.   This agrees with the practice of the ophthalmists and of the opticians, who always describe a converging lens as positive.   *A positive lens is one which imprints a positive curvature upon a plane wave which traverses it.*

The *unit of curvature*, whether of the wave-surface itself or of the surface of any mirror or lens, will be taken so as to accord with modern ophthalmic and optical practice as *the dioptrie ;* that is to say, the curvature of a circle of one metre radius will be taken as unity.   The dioptrie, originally proposed by Monoyer as the unit of focal power of a lens, was formally adopted in 1875 by the International Medical Congress at Brussels, and its great convenience has led to its universal adoption for the enumeration of the focal powers of lenses.   That lens which has a focal length of one metre is said to have a focal power of one dioptrie.   In other words, such a lens prints a curvature of one dioptrie upon a plane wave which is incident upon it.   For the present proposal to extend the use of the term from focal powers (*i.e.* imprinted wave-curvatures) to the curvatures of curved surfaces in general, the writer is responsible.

Notation

In adopting a notation which embodies the new method it is obviously advisable to choose one which lends itself most readily to the existing and accepted notations. In the great majority of books on optics, the recognised symbol for focal length is f; that for radius of curvature r. And in the Cambridge text-books for many years the distances from lens or mirror of the point-object and the point-image have respectively been designated by the letters u and v. Now it is the reciprocals of these which occur in the expressions for the curvatures of surfaces or of waves. The symbols adopted respectively for the four reciprocals are accordingly F, R, U, and V. The accepted symbol for the index of refraction is the Greek letter μ; for the velocity-constant, which is its reciprocal, we take the letter h. The following is a tabular statement of the symbols and their meanings:—

Symbol.	Meaning.	Equivalent in Current Notation.
F	Focal curvature, or Focal power of lens or mirror (= dioptries, if metre is taken as unit of length)	$\dfrac{1}{f}$
R	Curvature of Surface	$\dfrac{1}{r}$
U	Curvature of Incident wave ; i.e. curvature which it has acquired by having travelled from point of origin ("incident focus") to incidence	$\dfrac{1}{u}$
V	Curvature of Resultant wave ; i.e. curvature with which wave emerges from the lens	$\dfrac{1}{v}$
h	Velocity-constant of medium ; i.e. velocity of light in that medium compared with velocity in air taken as unity .	$\dfrac{1}{\mu}$

Expansion of Curvatures

If the curvature R of a wave at any point is known it is easy to calculate the curvature at any other point at distance d farther from or nearer to the centre, the formula for the new curvature R' being as follows :—

$$R' = R \frac{1}{1 \pm Rd} \qquad . \qquad . \qquad [1]$$

The $+$ sign must be taken where the new point is farther from the centre than the point for which the curvature R is specified ; the $-$ sign when it is nearer the centre. This proposition is of use in dealing with thick lenses, and with thin lenses at a given distance apart.

Refraction Formulæ

As a preliminary to lens formulæ, it is convenient to consider certain cases of refraction.

Consider a retarding medium, such as glass, bounded on the left (Fig. 39) by a plane surface SS'. Let P be a

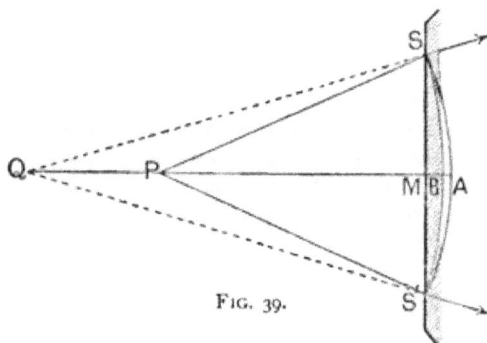

Fig. 39.

source of waves incident on the surface, PM being a line perpendicular to SS'. The wave-fronts, at successive small intervals of time, are represented by arcs of circles. At a certain moment the wave, had it been going on in air, would have had for its surface the position SAS'; the curvature being measured by the sagitta AM. The medium, however, retards the wave, and it will only have gone as far as B instead of penetrating to A; B being a point such that $BM = h \cdot AM$, where h is the velocity-constant of the medium into which the wave enters. The curvature of the wave is flattened as the result of the retardation. Now draw a circle through SBS', and find its centre Q. To a first degree of approximation the arc SBS' represents the retarded wave-front, the set of wave-fronts

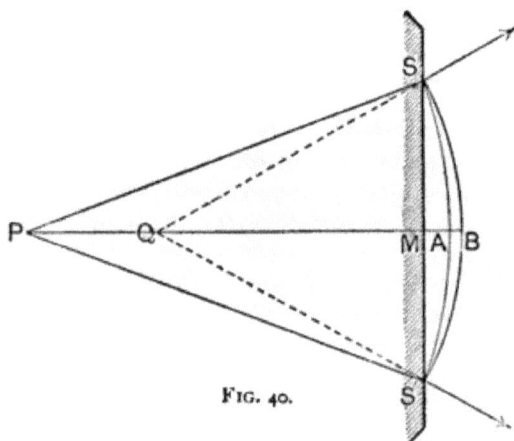

FIG. 40.

from B onwards being represented by the series of arcs drawn from Q as centre. An eye situated in the medium on the right of SS' will perceive the waves as though coming from Q, the (virtual) point-image of P. Accurately the wave-fronts should be hyperbolic arcs, but if SS' is small relatively to PM the circular arcs are adequate. Now $AM = U$, and $BM = V$. Hence the action of the plane surface upon the curvature (in this case a divergivity) of the incident wave is given by the formula

$$V = hU \quad . \qquad . \qquad . \qquad [2]$$

In the above case, which should be compared with Fig. 19, p. 34, the wave had a negative curvature. If the

entrant wave has a positive curvature or convergence such
as would cause it to march to a point P to the right in the
air, a similar set of considerations will readily show that if
on entering the flat surface of a more retarding medium its
curvature is flattened, it will march to a focus farther to the
right, the ratio of the original and the acquired curvatures
being, as before, dependent simply on the relative velocities ;
and formula [2] above still holds good.

Consider next the wave emerging (Fig. 40) into air from
a point P, situated in the retarding medium whose velocity-
constant is h. Had the wave been going on wholly through
the denser medium, the wave-front would have been at
SAS' ; but, being accelerated on emergence into air, it
reaches B instead of A. The new curve SBS' has Q for its
centre ; that is to say, the wave emerges from Q as a virtual
focus, its curvature being augmented. The sagitta BM is
greater than AM in the ratio of 1 to h. Hence in this
case the formula is

$$V = \frac{1}{h} U \quad . \qquad . \qquad . \qquad [3]$$

The case of an emergent wave of positive curvature
leads to the same formula. In the case of either positive
or negative initial curvature, emergence from the retarding
medium through the plane surface into air augments the
curvature.

If a plane wave travelling in air meets a bulging surface
of a more retarding medium such as glass, the portion of
the advancing wave which first meets the surface is retarded, so that the wave front receives a depression, and
hence on entering the second medium marches converging
toward a focus. The relation between the impressed focal
curvature and the curvature (R) of the surface is given by
the formula

$$F = R(1 - h) \qquad . \qquad . \qquad [4]$$

It will be noted that if the curvature of the surface is
positive (*i.e.* bulging toward the source of light), the impressed focal curvature is also positive. The formula,
therefore, is the same for entrant plane-waves whether the

surface be convex or concave, the sign of F following the sign of R. For the case of any two media having respective velocity-constants h_1 and h_2, the formula becomes

$$F = R\frac{h_1 - h_2}{h_1} \qquad . \qquad . \qquad [5]$$

A plane wave traversing a medium with velocity h and emerging through a curved surface into air has a curvature imprinted upon it that is of opposite sign to that of the surface itself. If the wave travelling to the right emerges through a (hollow) surface whose centre of curvature lies to the right, the acquired focal curvature will have its centre to the left, or will be negative; and its relation to the curvature (R) of the surface is given by the rule

$$F = R\left(\frac{h - 1}{h}\right) \qquad . \qquad . \qquad [6]$$

As before, for any two media having respective velocity-constants h_1 and h_2, the formula becomes

$$F = R\frac{h_1 - h_2}{h_1}, \qquad . \qquad . \qquad [5\ bis]$$

which, in the present case where $h_1 < h_2$, will give F of opposite sign to R.

The cases in which a wave possessing initial curvature passes through a curved surface and acquires a resultant curvature may be dealt with, apart from any further geometrical constructions, by applying the principle of superposition of curvatures. Thus, take the case of a wave possessing initial curvature U entering from air into a medium having velocity-constant h, and so curved that the focal power of the curved surface is F. Then, as the wave enters the surface of the medium two effects will occur: its initial curvature will be altered in the ratio of the velocities, and there will be superposed upon it the focal curvature of the surface; or, in symbols,

$$F_1 = hU + F_1 \qquad . \qquad [7]$$

For an emergent wave, possessing initial curvature U in the medium, the formula will be

$$V_2 = \frac{1}{h}U + F_2 \qquad . \qquad . \qquad [8]$$

Or, for the case of a wave passing from a medium of velocity-constant h_1 to another of velocity-constant h_2, the formula will be

$$V = \frac{h_2}{h_1}U + F \qquad . \qquad . \qquad [9]$$

It is easy, however, to prove any one of the several cases that may arise, without in this way relying upon the principle of superposition.

Lens Formulæ

In the case of a lens, the curvature F_1 imprinted on a plane wave by entrance at the first surface may be regarded as an initial curvature of the wave which emerges through the second surface. Emergence into air will, as shown above, alter the curvature by augmenting it in the ratio of 1 to h, and superpose upon it the focal curvature F_2 due to the second surface. Hence the whole resultant curvature F imprinted by a *thin* lens on the plane wave will be

$$F = \frac{1}{h}F_1 + F_2$$

But

$$F_1 = R_1(1 - h),$$

and

$$F_2 = -R_2\left(\frac{1-h}{h}\right);$$

whence

$$F = R_1\frac{1-h}{h} - R'_2\frac{1-h}{h},$$

or

$$F = (R_1 - R_2)\frac{1-h}{h} \qquad . \qquad . \qquad [10]$$

F

This formula may be compared with that in the current notation,

$$\frac{1}{f} = \left\{ \frac{1}{r_1} - \frac{1}{r_2} \right\} (\mu - 1).$$

Fig. 20 (p. 36), gives an illustration, in which however R_1 is zero, as the first face of the lens is flat.

In the case of a lens composed of a medium h_2, lying between two other media h_1 and h_3, the formula becomes

$$F = \frac{1}{h_1 h_2} \{ R_1(h_1 - h_2)h_2 + R_2(h_2 - h_3)h_1 \} \quad . \quad [11]$$

The general formula [10] for the power of any lens consists of two factors—one depending solely on the shape of the lens, the other upon its material. The latter factor, $\frac{1-h}{h}$, or $\mu - 1$, is a mere numeric; whilst the former, being the difference of two curvatures, is itself a curvature. If the curvature thus determined by shape solely is expressed in dioptries, then, on multiplying by the numeric which depends on the nature of the material, the resultant power of the lens will also be expressed directly in dioptries. In the optician's "lens-measurer" (p. 58) the dial readings are already corrected by being multiplied by this numeric, thus obviating calculation.

If the lens has thickness d, the rule for expansion of curvature at end of § 4 above at once gives

$$F = F_2 + \frac{1}{h} F_1 \frac{1}{1 \pm F_1 d} \quad . \quad . \quad [12]$$

or

$$F = \left\{ R_1 \frac{1}{1 \pm R_1(1-h)d} - R_2 \right\} \frac{1-h}{h} \quad . \quad [13]$$

Universal Formula for Lenses

The principle of superposition at once gives the universal formula for all lenses bounded by identical media on the two sides :—

$$V = U + F; \quad . \quad . \quad [14]$$

or, in words, *the resultant curvature is the algebraic sum of the initial curvature and the impressed curvature.* This may again be compared with the formula in current notation :

$$\frac{1}{v} = \frac{1}{f} - \frac{1}{u}.$$

The difference in sign attributed to the term $\frac{1}{u}$ arises from conventions adopted in the two systems.

Formula for Two Thin Lenses at a Distance Apart

The principle of expansion of curvature at once gives us as the equivalent focal power,

$$F = F_2 + F_1 \frac{1}{1 + F_1 d} \qquad . \qquad . \qquad [15]$$

where F_1 and F_2 are the focal powers of the first and second lenses, and d the distance between them. F will be in *dioptries* if F_1 and F_2 are in *dioptries* and d in metric units. If the two thin lenses are close together, the resultant power is simply the algebraic sum of the powers of the separate lenses. One simply adds the *dioptries* of the separate lenses to find the resultant *dioptries*.

Reflexion Formulæ

The plane mirror (Fig. 41) has surface SMS'. The incident wave would have had front SAS' at a certain instant had its path lain wholly in air. The central portion of the wave, which would have reached A, travels backwards to B, an equal distance, in the same time. The sagitta BM of the resultant curvature is equal to and of opposite sign to the sagitta AM of the initial curvature ; or

$$V = -U \qquad . \qquad . \qquad . \qquad [16]$$

There are two cases, equally simple, of convex and concave mirrors. These are separately shown in Figs. 13 and 14 (pp. 25, 26), in both of which the incident waves are plane. Consider (Fig. 42) a plane wave which at a certain instant would have arrived at SAS' had its path lain wholly in air. The central portion of the wave has, however, struck at M, and marches backwards to B in same time as it would have taken to reach A.   Hence

$$BM = AM,$$

or

$$BA = 2AM.$$

But AM measures the curvature of the mirror, whilst BA

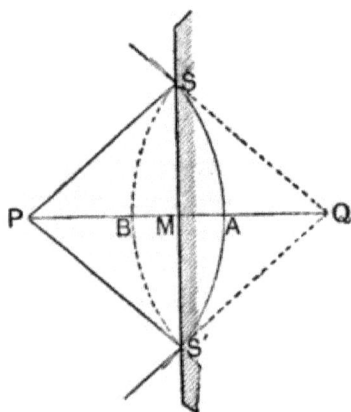

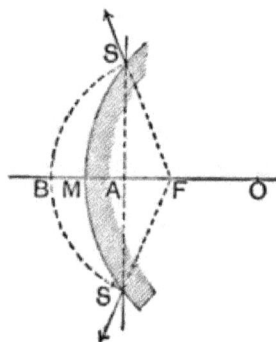

Fig. 41.   Fig. 42.

measures the curvature impressed on the plane wave. Hence

$$F = 2R \quad . \qquad . \qquad . \qquad [17]$$

In Cambridge notation

$$\frac{1}{f} = \frac{2}{r},$$

or

$$f = \frac{r}{2}.$$

It is equally easy to establish the formula for the action of a curved mirror on a curved wave. The principle of

superposition at once leads to a general formula, expressing
the sum of the two actions of the mirror on the wave ; it
reverses its initial curvature, and then imprints a focal
curvature upon it.   In symbols,

$$V = - U + F \qquad . \qquad . \qquad [18]$$

The application of wave principles to find the direction
of a refracted beam is best handled by Ampère's modification of Huygens's construction, as in Fig. 43.   In that
figure the dispersion produced by the difference between
the velocities of light of different colours is also illustrated.

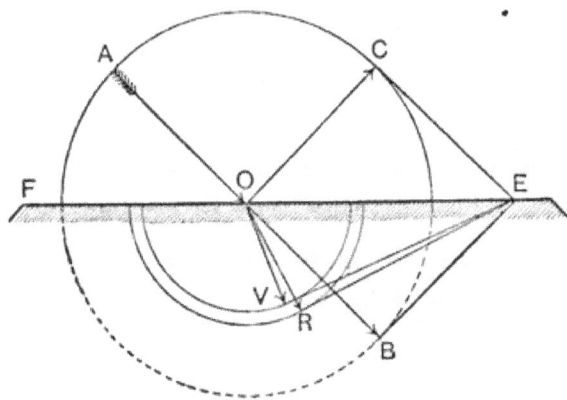

Fig. 43.

During the time that light (of any colour) would travel 1
foot in air, red light would travel about $7\frac{1}{2}$ inches in glass,
and violet light about 7 inches in glass.   The velocity-constant for glass of this kind is 0·625 for red waves, and
0·583 for violet waves.   Let EF be the surface of glass
(Fig. 43) at which the wave is entering.   It marches in the
direction AO.   Consider the portion of wave-front that
arrives at O.   If it is regarded as setting up a new set of
waves, these will spread in circles according to the velocity.
Therefore around O as centre draw a number of circles, one
with unit radius to show how the wave would have spread
in a given time had it spread all round in air, one (a semi-

circle only) with radius 0·625 to show how far the red wave would have penetrated in glass, one (also a semicircle) with radius 0·583 to show how far the violet wave would have travelled in glass in the same time. Now produce OA to B, and at B draw a tangent meeting the surface of the glass at E. From E now draw as many tangents as you can to the circles, to represent the wave-fronts. EC will be the wave-front of that part of the light which is reflected back in the direction OC ; ER will be the wave-front of the red light refracted down along the direction OR ; and EV will be the wave-front of the violet light refracted down the direction OV.

LECTURE II

THE VISIBLE SPECTRUM AND THE EYE

Colour and wave-length—Rainbow tints—The spectrum of visible
colours—Spectrum made by prism—Spectrum made by grating
—Composition of white light—Experiments on mixing colours
—Analysis of colours—Blue and yellow mixed make white, not
green — Complementary tints — Contrast tints produced by
fatigue of eye—Other effects of persistence of vision—Zoetrope
—Animatograph

WAVES of light are not all of the same wave-length. The
difference of size makes itself known to our eyes as
colour. Just as the sounds of different wave-lengths
produce in our ears perceptible differences in pitch, so
the lights of different wave-lengths produce in our eyes
different sensations, which we call colour. Any simple
kind of light—I am not speaking here of mixtures—can
be described in two ways; either (1) by stating the
colour-sensation which it produces on the eye, or (2)
more accurately, by stating what its wave-length or the
frequency of its vibrations is. To ascertain the wave-
length of any particular kind of simple light may not
be a very easy matter, but when once it has been
measured, the statement of the wave-length is an
accurate description.

To begin then, here is a table in which I have set down, in their order according to wave-length, biggest first, the various kinds of simple light that are visible to the eye.

TABLE I.—COLOURS OF THE SPECTRUM

NAME OF COLOUR.	Wave-length in millionths of an inch.	Wave-length in millionths of a centimetre.
Extremest red . .	32·4	81·0
Red	26·0	65·0
Orange . . .	23·3	58·3
Yellow . . .	22·0	55·1
Green	20·5	51·2
Peacock . . .	19·0	47·5
Blue	18·0	44·9
Violet	16·0	40·0
Extreme violet . .	14·4	36·0

You will note that the red waves are about twenty-six millionths of an inch long (*i.e.* about $\frac{1}{39000}$ of an inch), while the violet waves are a little more than half as great, namely, sixteen millionths of an inch in wave-length (*i.e.* about $\frac{1}{60000}$ of an inch). All the other simple kinds of light are of intermediate size. You will note the names of the colours. In the list you will find neither white nor black, for white (as I shall presently show you) is a mixture of all these simple colours, and black is simply the absence of all light—a mere darkness.

Now this set of colours can be produced naturally in their proper order in several different ways. The simplest way is to take some white light which contains all these colours mixed up together, and sort them out. But how? That is what I want you to understand. In nature we find them sorted out in the rainbow, where these tints stand side by side. Can we make an artificial rainbow? How is a rainbow made? Of the smiles of Heaven commingled with the tears of Earth, if we believe the poets.[1] Of sunlight and raindrops—(is it not?)—which refract the light, and in refracting it sort out the different kinds of light, and display them in their proper order. Perhaps that is a very incomplete description of the operation of building a rainbow, but it is good enough to give us a hint towards experiments.

Here is an optical lantern, with an electric arc-lamp inside, a sort of miniature sun to give us white beams of light. We let the light pass out in a fine straight beam, and in that beam we place—to serve as a sort of magnified raindrop—this sphere of water contained in a thin shell of glass. See the bow which it casts back upon the whitened screen. You can recognise the usual tints, though they are not so brilliant as in the natural bow.

But having got our clue to experiment, let us go on farther. Try instead of the bulb a three-cornered bottle full of water. We have now no bow. The beam of light is abruptly turned upward into a new direction, and falls upon the wall or ceiling. But, though we have lost the shape of the arch we have gained in the development of the rainbow hues. We have now a brilliantly coloured, though rather nebulous, colour-patch. Try again, and

---

[1] "We are like Evening Rainbows, that at once shine and Weep—things made up of reflected splendor and our own Tears."— *S. T. Coleridge.*

this time try the effect of varying the liquid. Here is a three-cornered bottle full of turpentine. The angular deviation of the colour-patch has become greater, but so has the breadth of our set of tints. Try oil of cinnamon, it is still better. Try bisulphide of carbon, still more brilliant though still fuzzy at the edges. Naturally, one begins to think that if a transparent three-cornered bottle full of liquid will thus display rainbow effects, a three-cornered piece of transparent glass ought to do the same. So it does: and so we have arrived at the use of the well-known glass prism to produce a *spectrum* of colours. The word *spectrum* means simply "an appearance": in this case an appearance of colours—the colours sorted out in their order. To emphasise the fact that the spectrum is in this case produced by use of a prism, it is sometimes called the "prismatic spectrum." In all cases you will have noticed that the order of the colours is the same, and that the red light is always refracted least, and the violet light refracted most. If the refractions of these colours were equal, the prism would not separate them. The difference of the refractions between the most-refracted (violet) and least-refracted (red), of the visible kinds of light, is sometimes called "the *dispersion*" of the prism.

We have now got to the stage of Newton's researches, but there is one further improvement to make, which was indeed tried by him. Let us try the effect of altering the arrangement of our beam of light. You see we have been using a beam streaming out through a round hole ; when nothing is interposed it falls in a round spot against the wall. Newton used sunlight streaming

through a hole in a shutter.   Well, let us try the effect
of using holes of different .izes and shapes.   And, while
we are about it, let us try the effect of focusing on the
wall the image of the aperture—by interposing a positive
lens—so as to work with a well-defined spot of light
instead of a fuzzy patch.

We begin by using round holes of different sizes,
which we can try one after the other.   Now, interpose
the prism—the best one of those yet tried—in the path
of the light.   You see that when the aperture used is a
large circular hole, the colours overlap much near the
middle and give a mixed effect.   Whereas when we use
a smaller hole, though we have less total light, the
colours are more intense, simply because they overlap
less.   Well, then, let us take the hint, and substitute for
the small round hole a narrow slit.   By employing a slit
with movable jaws (like a parallel ruler) we can adjust
it to be as wide or as narrow as we like.   Again, we find
that if the slit is too wide the colours overlap, while
with a narrow slit the tints are more intense.

Our successive improvements have then led us to
the following combination : a slit to limit our beam, a
lens to focus the image of the slit as a fine white line
on the screen, and a prism, which, in refracting the light
of the lamp, also splits it up (Fig. 44) into the various
colours of which it is compounded.

Perhaps you think I am assuming things not yet
proved to describe the action of the prism as splitting
up the light of the lamp into the colours of which it is
compounded.   Well, I admit, the phrase "splitting it
up" is not the best that might be selected; "sorting it

out " would be a better phrase. But each of these
phrases carries in its use the assumption that the white
light is a mixture that can be split up or sorted out into
simpler constituents. That is precisely Newton's great
discovery. White light, supposed down to that time
to be itself a simple thing, was found and proved by
him to be a mixture. The prism added nothing to the
white light, it simply spread out the constituents in their

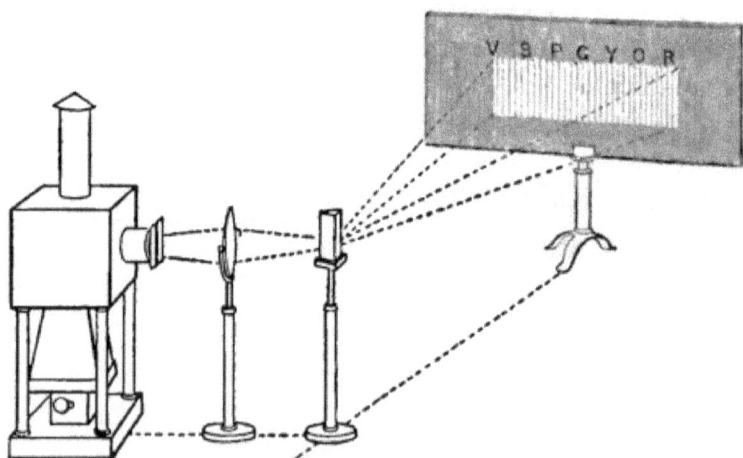

FIG. 44

natural order. More than a hundred years afterwards the
great poet and dramatist Goethe—"master of those who
know "—fought against this idea, and threw the whole
weight of his genius to demonstrate, in his *Farbenlehre*,
the erroneous nature of Newton's views. According to
him the prism does not merely spread out the simple
constituents of white light: it takes simple white light
and adds something to it which gives it a tint of one
sort or another. But it was in vain. Beautiful as many
of Goethe's experimental researches were, his theory

died of inanition.   To-day not a single scientific man, even in Germany, holds Goethe's theory of optics; though his fame as a poet stands immortal.

Before we follow the quest of those other experiments by which Newton's theory of the compound nature of white light is established, let me show you a second method of spreading out white light into a spectrum of colours.   In this case I use no prism ; and the effect will not be produced by refraction through any transparent solid or liquid.   Instead, I employ the little instrument which I hold in my hand.   It is called a "diffraction grating."   It is simply a polished mirror of hard bronze, a little more than two inches wide, across the surface of which there have been ruled with a diamond a large number of parallel and equidistant scratches. You may think it odd to call a scratched mirror a "grating."   But the fact is that the properties it possesses were originally discovered by the use of gratings made of fine wires.   It would be quite impossible, however, to make a grating with wires as fine as these scratches.   When you want to produce a perfect diffraction grating there is nothing for it but to rule diamond scratches ; and they must be ruled by machinery of the utmost precision.   Over the face of this little mirror there have been ruled about 30,000 parallel lines, and not one of them is a millionth of an inch out of its proper place.   It was ruled at Baltimore on Professor Rowland's machine.   The exact number of lines is 14,400 side by side to the inch.

I set up the grating so that the light of my lantern, issuing through the slit, falls upon it, and you see the spectrum that it casts upon the wall. This is not a *prismatic spectrum*, for there is no prism. It is a *diffraction spectrum ;* and that is not quite the same thing. As a matter of fact the grating, as you see, casts on the wall a whole series of spectra. It reflects back centrally a white image of the slit. Right and left we have on each side a bright spectrum with all the colours. Then, still farther away on each side, a rather longer and nearly equally brilliant spectrum of the second order ; while, more dimly, and slightly overlapping one another, we have spectra of the third and fourth orders. We will deal only, however, with the first bright spectrum. There are our rainbow tints in their order as before. But note that now it is the red light that seems to have been turned most aside, and the violet light which is least. Note, further, that while the order of the colours between red and violet is the same as in the prismatic spectrum, the spacing of them is not the same. In the prismatic spectrum the orange is huddled up toward the red, and the yellow toward the orange ; while the violet and blue are highly elongated. In the diffraction spectrum the red end is not squeezed together unduly, nor the violet end unduly drawn out.

Time will not allow me [1] to dwell on the reasons for these differences. Suffice it to say that they depend upon the wave-lengths of the different kinds of light, and their relations on the one hand to the size of the molecules of the refracting prism, and on the other hand to the width of the bars of the grating.

Incidentally you may be interested in knowing that

[1] See Appendix, p. 100.

this property of diffraction, which belongs to the surface
that has thus been covered with parallel scratches, can
be transferred from the grating to another surface by
merely taking a cast. Here is a cast made in gutta-
percha [1] from the grating; it is itself a grating. Like the
bronze original it glitters with rainbow tints, and will
throw a set of spectra on the wall. Mother-of-pearl
glitters with rainbow-tints for precisely the same general
reason, it possesses naturally a structure of fine striations
or ridges which produce (rather irregularly) diffraction.
But as the ridges are not quite equidistant, the tints are
never pure. But, do you know, if you will take with
sealing wax—black wax is best—an impression from a
piece of mother-of-pearl, you will find it glitter just as
the mother-of-pearl does.

Whichever of these two means we use—prism or
grating—of producing a spectrum, you will note that
what we do is to sort out the mixture into its constituents; we analyse the light. Presently we shall be
able to make use of this sorting process to discover what
some of the compound colours are made up of. But in
the meantime we will return to the prismatic method to
show some further experiments.

To produce a good arched rainbow artificially, but in
all the splendour of the natural colours, I have recourse
to a specially constructed compound conical prism. A
glass cone of light crown glass is mounted with its point
turned inward (Fig. 45), within a hollow truncated cone
of glass, the face of which is closed with a glass plate.
The annular space is filled with a highly refracting liquid,

[1] Made by Mr. E. Rousseau; see footnote, p. 31 *ante*.

cinnamic ether.[1] An annular slit in a plate of tin-foil is
fixed against the end of the cone ; and the whole prism

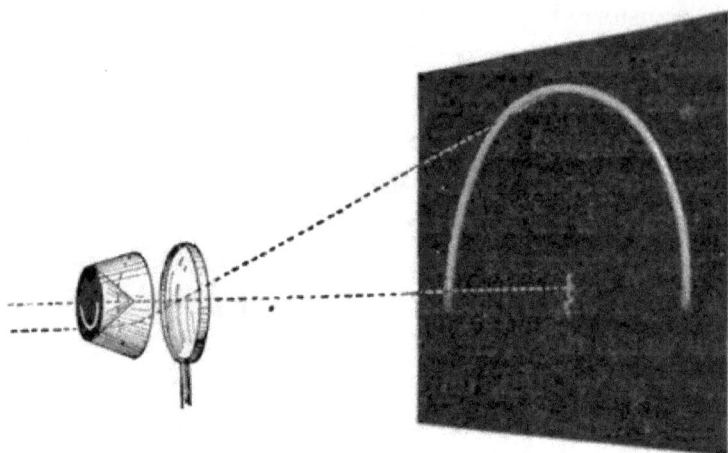

FIG. 45.

is placed in a nearly parallel beam of light issuing from
the lantern. Beyond it is a lens to focus the light.

[1] This liquid is excellent for direct-vision prisms, for it has
exactly the same mean refractive index as one kind of the light
crown glasses made at Jena. Figs. 46 and 47 depict the direct
vision prism with parallel end-faces designed by the author in 1889,
and constructed by Messrs. R. and J. Beck. A prism A of this glass

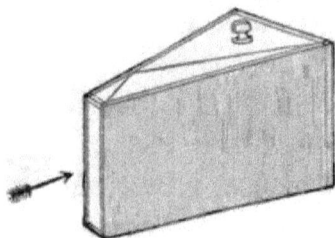

FIG. 46.

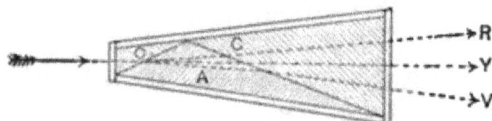

FIG. 47.

with a refracting angle of 135° is immersed in a glass cell filled with
cinnamic ether. Yellow light, as shown in Fig. 47, goes straight
through, while red light is thrown to one side and violet to the other.
This prism is very suitable for projecting the spectrum on the screen.

Thus we project upon the white screen in almost exactly true proportions a rainbow. Note the order of the colours, red along the outer edge, then orange, a trace of yellow, green, peacock, blue, and lastly violet along the inner edge. This is the correct order as in the natural rainbow. But you often see it incorrectly depicted by artists—they put the colours in the wrong order, or with the red along the inner edge and the violet along the outer.

My assistant will now give us upon the screen the bright spectrum which we saw before, in order that we may study the effects of the different kinds of light on coloured stuffs. Here is a piece of blue drapery, and here a piece of scarlet. What is the effect of putting these into the spectrum, first into one kind of light, and then into another? If we put the blue stuff into the red or orange or yellow light it looks simply black. But in the blue part of the spectrum it looks blue, in the violet part it looks violet, in the green part it looks green. Clearly the surface of it is incapable of reflecting back either red, orange, or yellow, while it is capable of reflecting back green, blue, and violet. In fact, when ordinary daylight falls on it, it absorbs some of the waves and destroys them, while it reflects back to our eyes some others of the waves, and gives us on the whole a blue effect from the green, blue and violet waves mixed together and reflected back. The red stuff, when placed in the red part of the spectrum, looks red, and in the orange and yellow parts it looks dull orange and dull yellow; while in all the other parts of the spectrum it looks black. This red stuff then absorbs

and suppresses all the violet, blue, peacock, and green
rays, and reflects back only those at the red end of the
spectrum. But, of course, it would only look red if
there was some red light present. And the blue stuff
would only look blue if there was some blue light present. The colour is really not in the stuff, it is in the
light that the stuff reflects.

To prove this let us see how these red and blue stuffs
appear when we shine upon them
some light that has neither red, nor
blue, nor green, nor violet in it, but
has yellow only. The monochromatic lamp which Professor Tyndall
used to employ here (Fig. 48) has
been lit. It consists of an atmospheric gas-burner, into the dim
flame of which salt is projected,[1]
making a splendid yellow flame
devoid of every other kind of light.
I hold these blue and red stuffs in
the light of the yellow flame. The
one appears simply black, the other
a dull gray. A set of stripes of
gay colours painted upon a board
appear simply dull grays and blacks, except the yellow
stripe, which seems brighter than all the others. Even
gaily-coloured flowers seem merely black or gray ; while

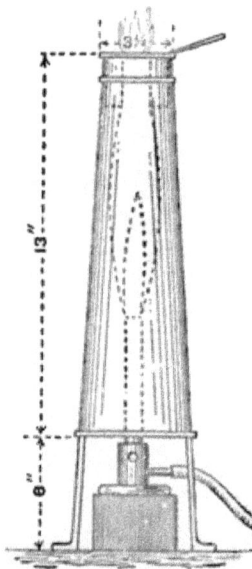

Fig. 48.

---

[1] The salt is contained in an annular pan at the top of an external
tapering chimney of sheet iron. This annular pan has a gauze
bottom, through which on tapping the chimney the salt falls in fine
powder into the flame.

the complexion of the human countenance appears simply ghastly.

Now if Newton's view is correct that white light consists of the lights of various colours mixed up together, it ought to be possible to make white light by taking lights of all the various colours and mixing them together. Do not try to mix together pigments out of your paint box—they won't make white paint when mixed. That is because pigments are not lights —they are darknesses rather than lights. Think for an instant what you do when you want to paint a card crimson. You take a piece of white card, and paint over it a pigment which darkens it, so that it sends back to your eyes crimson only, and absorbs the other parts of the white light. No, you must not mix pigments—you must mix lights, and mix them in the correct proportions.

Now there are several ways of doing this; and first of them we will take the spectrum colours and recombine them to produce white light. We take the spectrum light as it issues obliquely from the prism, and reflect it upon the screen with a piece of silvered mirror-glass. By simply waggling the mirror-glass upon its stand, we cause the spectrum to oscillate rapidly across the screen. The colours then all blend by rapid superposition, and we obtain a white band bordered by colour only at the ends.

Another way to recombine the spectrum is to employ a cylindrical lens (Fig. 33, p. 49), so placed in the path of the diverging coloured rays that it collects them back to a focus on the screen, and gives us back the image of our slit as a white streak.

An independent plan, suggested by Newton, is to paint upon a circular card[1] (Fig. 49), in narrow sectors, the various tints in proportions ascertained by experiment to give the best result; and then, putting this upon a small whirling-table, spin it round so fast that the colours all blend in the eye, giving, when well illuminated against a black background, the effect of white. A similar arrangement can be made for use in the lantern, the sector-disk being painted in transparent tints, or coloured by affixing narrow wedges of coloured transparent gelatine.

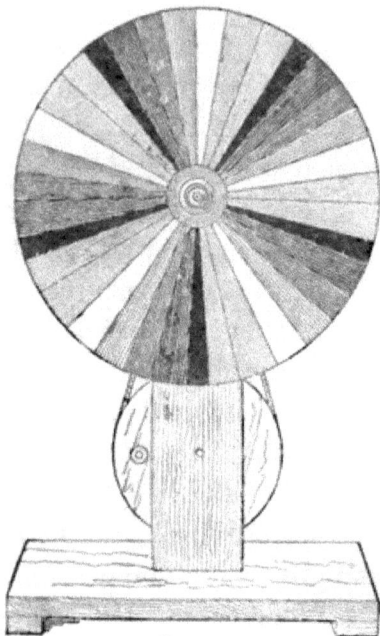

FIG. 49.

This method of colour-mixing by whirling round before the eye surfaces tinted with the colours desired to be mixed, is capable of extension to other cases. Suppose we wish, for example, to mix red and

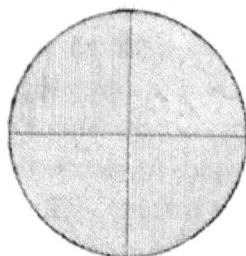

FIG. 50

green, or blue and orange together, we have only to paint

[1] The colour-whirler actually shown was lent by Messrs. Harvey and Peak, who use strips of brilliantly tinted paper pasted upon a card in such a way as to repeat the gamut of colours from red to violet five times around the circle. If the colours are thus repeated the card does not require to be whirled very fast to produce white.

a round disk (Fig. 50) with the colours desired to be mixed, either in semicircles, quadrants, or in any other desired proportion, and place them upon a whirling machine to see the effect.

The arrangement which I now show offers an improvement in several respects.   Upon the whirling-table is fixed a light cylinder of wood, which is slightly tapered toward the top, so that over it may easily be slipped a paper sleeve or tube upon which the colours are painted. I have here several of these paper tubes. The colours (in most cases coloured paper being cut to shape and pasted on) to be mixed are arranged in two sets of narrow

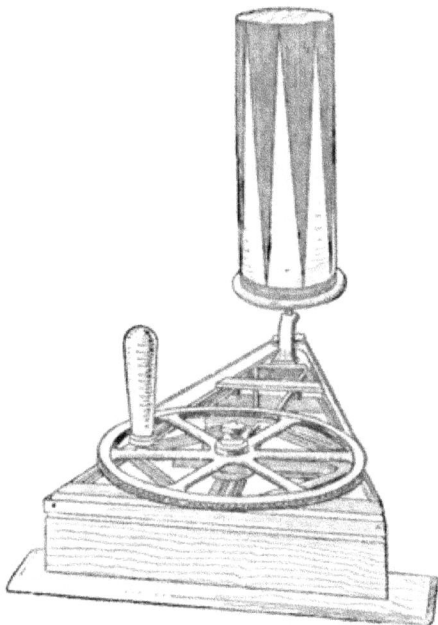

FIG. 51.

triangles, as shown in the figure.   When these are whirled, one gets combinations in all proportions. For instance, if red and green are the two colours chosen, one end of the revolving surface is full red, the other full green, and the colours gradually fade one into the other.   About the middle, one obtains a curious gray, which if seen by daylight looks rather greenish ; but by gas-light or lamp-light looks rather reddish owing to the greater relative prevalence of red waves in artificial light. This change of apparent tint is similar to that observed in the rare gem called the alexandrite,[1] which is green by day and deep red by night.

Returning from the operations of colour-mixing by rotation, I return to the property of the prism to analyse mixed lights by spreading out the constituent colours as a spectrum. Newton tried an experiment to see whether if you took light of one tint alone you could split it up still further by passing it through a second prism. I introduce across the path of the spectrum on its way to the screen a diaphragm of cardboard, having a narrow slit in it. I push it along so that the slit allows waves of but one particular colour—say green—to pass. Now, if I interpose beyond this slit a second prism, I find that it turns the beam of green light round at an angle, and widens it out a little more, but it does not split it up into any other colours: it is still a green beam. So it would be with any other. When once you have procured a simple tint by dispersing away the other colours to right and left, the prism effects no further analysis[2] of colour.

---

[1] A gem of the emerald species found in a mine in Siberia belonging to the Imperial Russian family.

[2] In this sense every pure spectrum tint is a primary tint, and the number of such tints incapable of further analysis is infinite. Each kind of light of a given wave-length is thus a simple tint. But the eye possesses three different sensations of colour, each of which is physiologically a primary sensation. These three primaries are, a *red*, a rather yellowish *green*, and a *blue-violet*. Any other tint than these excites more than one sensation. For instance, a pure spectrum yellow excites both the red and the green sensations; therefore yellow cannot be called truly a primary. In the same way peacock tint excites the green and the blue-violet sensations.

Now let us try a few experiments in the analysis of colours by the prism. There are many well-recognised tints, known by familiar names, which are not to be seen in the simple colours of the spectrum, for the simple reason that they are compound colours. In the spectrum there is no purple; for purple is a mixture of red from one end of the spectrum with violet or blue from the other end. Pink does not exist in the spectrum—for pink is red, diluted by admixture with white, that is to say, with a little of every other colour. Neither is there any chocolate colour, which is red or orange diluted with black, that is to say, a little red or orange spread where there is no light of any other colour. Buff, olive, russet, bistre, slate, and many other colours are also compounds. Well, whatever they are, the prism can analyse them. Here is a piece of gelatine, such as you may get off a Christmas cracker, stained a beautiful purple. Why does it look purple? What kinds of light does it actually allow to pass through it that it should look purple? I have merely to interpose it in the path of the white light for you to see the beautiful purple colour on the screen. Now placing the prism in front of it you see the purple spread out into its constitutents. There is red at one end; there are violet and blue at the other. But in between, where orange, yellow, green, and peacock colours should come, there is darkness. The purple stain in the gelatine cuts off all these and lets the others go by. Here is another piece of gelatine stained with magenta—you see it lets more red and a little violet and blue go through. Here is a small glass tank containing the pale purple liquid

known as Condy's fluid (permanganate of potash); on
interposing it across the beam of light through the prism
you see (Fig. 52) that it cuts off the yellow and greenish
yellow, but transmits red and orange at one end of the
spectrum, and at the other violet, blue, peacock, and
some green.   Here are some coloured liquids in bottles
(Fig. 53); red liquid (amyl alcohol dyed with aniline-
red) floating on the top of a green liquid (cupric

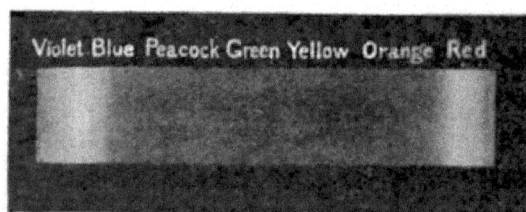

Violet Blue Peacock Green Yellow Orange Red

FIG. 52.   FIG. 53.

chloride dissolved in dilute hydrochloric acid) without
mixing.   The red—as you see when I expose it to
analysis in the spectrum—is a good red—it cuts off every
tint except red.   The green is also a fairly good green—
it cuts off everything except green, peacock, and a trace
of blue.   What will happen if I now shake up the two
solutions and mix them ? I obtain a mixed liquid[1] that
cuts off everything, and is simply black.   Many other
experiments of an instructive kind may be tried with

[1] The liquids named possess the very convenient property of
separating from one another in a very few minutes. In preparing
the experiment a little trouble and care is required to get the solutions to balance. By adding first a little of the red liquid and then
a little green as may be required, and trying the effect of shaking
up, the liquids may be adjusted. Various other colour-combinations
are possible in this way.

coloured liquids, their apparent tints depending on the kinds of light they absorb and transmit respectively. Any liquid which merely absorbs green will look reddish, since in the balance of colours it transmits, the complementary red will preponderate. Similarly any liquid (or glass) which merely absorbs the blue part of the spectrum will look yellow. It is even possible to find a liquid,[1] which though it looks yellow to the eye really transmits nothing but green and orange, which when mixed have the same effect on the eye as yellow. This proves that the sensation of yellow, though it may be excited by a simple spectrum tint of a particular wave-length, can also be excited by a mixture of other tints, and is therefore not a primary colour-sensation as red, green, and blue-violet are.

And this brings me to another point, viz. that while yellow light can be thus made by mixing together orange and green lights, it is found to be absolutely impossible to produce green[2] by mixing together any two other pure lights. Blue light and yellow light, as remarked above, do not when mixed produce green, but white. This is so fundamental a matter that it is worth while to illustrate it by further experiment.

My assistants have two lanterns. From each of them there is now thrown upon the screen a round white disk of light. In front of one lantern is interposed a film of blue gelatine—and that disk turns unmistak-

---

[1] Mixed solutions of chromic chloride and potassium bichromate.

[2] Just as also it is impossible to produce red light by mixture of any other two simple lights, or to produce blue-violet by admixture of any other two simple lights. These three—red, green, and blue-violet being the three primary colour sensations.

ably blue. In front of the other lantern is interposed a film of yellow gelatine, and the second disk of light on the screen becomes bright yellow. Now one of the lanterns is turned a little aslant so as to make one of the disks overlap the other (Fig. 54). Where they over-

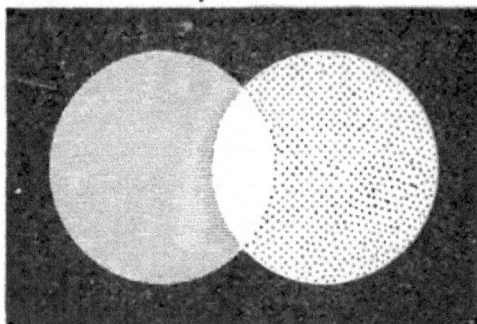

FIG. 54

lap and the lights mix we have—not green—but white!

I put in the lantern a colour-whirler, having a disk covered over half with blue and half with yellow gelatine, and on whirling it round the blue and yellow mix, and make white.

Here is an experiment that any boy might make at home. A cardboard disk is divided into twelve sectors, six of which are covered with blue paper, and the alternate six with yellow (Fig. 55). I put a pin through the centre, and spin it round by hand—and behold blue and yellow are mixed, and make white.

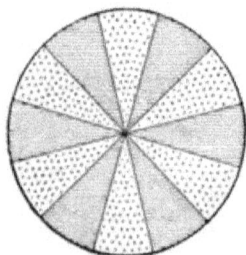

FIG. 55.

We give the name of "complementary" tints to any

pair of tints which thus mixed together make white. That is to say, if any two tints are so related that each contains the constituents that are wanting in the other, then we describe them as the complement one of the other. Here is a table of some tints which experience shows to be complementary one to the other:—

TABLE II.—COMPLEMENTARY TINTS

Crimson is complementary to	.	Moss green
Scarlet	,,	.	Peacock
Orange	,,	.	Turquoise
Yellow	,,	.	Blue
Primrose	,,	.	Violet
Green-yellow	,,	.	Purple

There are other cases also not set down on the list. Now seeing that the sensation of white is not excited unless all three primary sensations [1] (red, green, violet), are stimulated at the same time, it is clear that when two colours are found that are complementary to one another, by no possibility can both be primary colours. One may be, but in that case the other will be a mixture of the two other primaries. If primary red is one of the two complementary tints the other will be a bluish-green or peacock colour made up of primary green and primary blue-violet mixed.

Probably most of you are aware of the subjective colours that are seen on closing the eyes after looking

[1] The red that is primary is a full red. The green that is primary a rather yellowish-green, the violet a rather bluish-violet.

at a bright light. These are connected with the fatigue
of the nerve, and with the residual nervous stimulation.
But closely connected with them are the "contrast"
colours that are seen on a gray background after the
eye has been fatigued by looking at any coloured object.
The tints of these "contrast" colours are approximately, though not accurately, the complementaries of
the respective colours that have excited them. Thus
after staring intently for
some time at a bright
green disk, the after-image
against a gray wall is of
a reddish tint. There are
many ways of showing
these tints. I will give
you as an example that
used here in this theatre
by the late Professor
Tyndall. Against the
white wall, half-lit with
daylight, I hold up on
the end of a stick a

Fig. 56.

cardboard disk about a foot in diameter covered
with bright blue paper (Fig. 56). The beams of an
electric lamp are directed upon it to make its tint more
brilliant. You must look at it fixedly while I count
thirty in a distinct voice. When I come to "thirty" I
will drop the disk; but you must continue to look at
the same region of the wall, where you will see—now
that I drop the disk,—a yellow image or ghost, of the
same size as the blue disk. In like manner if I hold up

a red disk you will, when your eye is fatigued, see a green
or peacock coloured ghost.   The reason of the contrast
tint is that if you have fatigued the eye, or any region
of the retina of the eye, with red waves, that region will
be less sensitive for red than it is for the other colours.
Hence if gray (*i.e.* diluted white) is presented in view,
the retina at the fatigued region is more sensitive to
all the other tints present than it is to red, and will
therefore on the whole receive an impression in which
green predominates.

Another way to see the contrast tints is to stretch
upon a ring of cardboard a sheet of semi-transparent

coloured tissue paper, and then
upon this as a background gum
a smaller ring of white cardboard.   Diffuse daylight should
be allowed to fall from the front
upon the white card, making it
gray.   By lights suitably placed
behind the coloured tissue is
lit up.   The eye therefore sees
the gray ring between an inner
and an outer circle of colour.

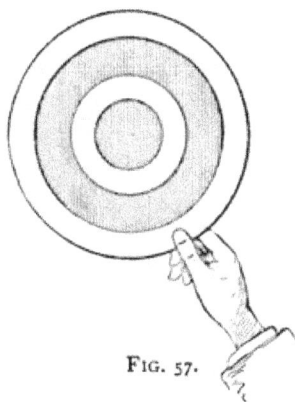

Fig. 57.

And after looking for a very few seconds will pronounce
the gray card to have become of a complementary tint.
Thus if the tissue paper is orange in hue the gray card
ring takes a bluish-peacock colour by contrast.

But the persistence of impressions in the eye is not
limited to phenomena of colour.   All ordinary visual
impressions last a perceptible time, the images of brightly
lighted objects, even when only viewed for a thousandth

of a second will take a whole tenth of a second to die away. If, then, you can present to the eye a second

FIG. 58

impression before the first one has died away, the effect is the same as though both had been present at one time. A familiar toy depending on this principle, and called the thaumatrope, consists of a bit of white card held between two strings on which it can be twirled. On one side there is painted say a horse ; on the other his rider. On blowing against the card to twirl it you see the rider (Fig. 58) mounted on his horse. We may try this experiment in a new way. On one side of a vertical card is painted in outline a birdcage. On the

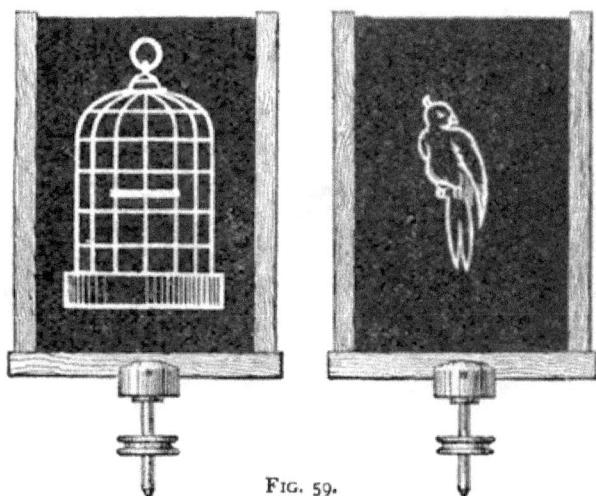

FIG. 59.

other side, a bird. By a band and pulley we spin the card rapidly ; and lo ! you see the bird within its cage.

I hold in the lant .n a small disk of sheet metal having a number of holes pierced in it ; giving on the screen a lot of bright points of light. Then on making this disk vibrate on the end of a spring, or rotate about a pin, each little white hole is transformed apparently into a luminous line, moving about on the screen.

Another optical illusion depending chiefly upon the persistence of vision is afforded by the strobic circles which I devised in 1877. On giving these black and white patterns a small "rinsing" motion, the circles and toothed wheels seem to rotate on their axis.

The latest of optical illusions and one not easy to explain,[1] is Benham's colour-top. A number of narrow black-lines are drawn as arcs of circles of various lengths, upon a white surface, half of which (Fig. 61) is coloured

FIG. 60.

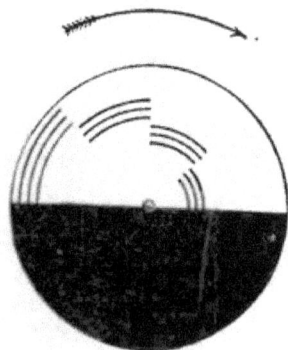

FIG. 61

black. On revolving this disk, and viewing it by a sufficiently strong light, the arcs of some of the circles appear coloured. The rotation must be neither too slow nor too quick. On reversing the rotation the order of the colours reverses. The effect appears to be due to the intermittent stimulation.

[1] See recent paper in *Proceedings of the Royal Society*, by Mr. Shelford Bidwell, F.R.S., whose explanation is that when the particular nerve-fibres which give the red sensation are excited at any part of the retina, the immediately adjacent parts of the same nerve-fibres are for a short period sympathetical'y affected, so that a red border seems for an instant to grow around the image of a white object suddenly seen. In the same way, when the image of a white object is suddenly cut off there is a sympathetic reaction giving a transient blue border around the disappearing image.

Another example of effects produced by persistence of the optical impressions in the eye is afforded by an old toy, the *zoetrope*, or wheel of life; in which the semblance of motion is given to pictures by causing the eye to catch sight, in rapid sequence, through moving slits, of a series of designs in which each differs slightly from the one preceding. Thus if you want to make the sails of a windmill seem to go round, the successive pictures must represent the sails as having turned round a little during the brief moment that elapses between each picture being glimpsed and the next being seen. These intervals must be less than a tenth of a second, so that the successive images may blend properly, and that the movement between each picture and the next may be small. Mr. Muybridge has very cleverly applied this method to the study of the movements of animals. Anschütz's moving pictures, illuminated by intermittent sparks, were the next improvement. And the latest triumph in this development of the subject has. been reached in the *animatograph*, which the inventor, Mr. R. Paul, has kindly consented to exhibit.

The animatograph pictures are photographed upon a travelling ribbon of transparent celluloid; the time which elapses between each picture being taken and the next being about one-fiftieth of a second. A scene lasting half a minute will, therefore, be represented by about 1500 pictures, all succeeding one another on a long ribbon. If these pictures are then passed in their proper order through a special lantern, with mechanism that will bring each picture up to the proper place between the lenses, hold it there an

H

Fig. 62.

instant, then snatch it away and put the next in its place, and so forth, the photograph projected on the screen will seem to move. You see in a street scene, for example, the carts and omnibuses going along; the horses lift their feet, the wheels roll round, foot passengers and policemen walk by. Everything goes on exactly as it did in the actual street. Or you see some children toddling beside a garden seat. A big dog comes up, and the boy jumps astride of him, but falls off (Fig. 62), and rises rubbing his bumps. Or a passenger steamer starts from Dover pier: you see her paddles revolve, the crowd on the pier wave farewells with handkerchiefs or hats, the steamer wheels round, you see the splash of foam, you note the rolling clouds of black smoke proceeding from her funnel, then she goes out of sight round the corner. The reality of the motions is so great that you feel as though you had veritably seen it all with your own eyes. And so you have. You have just as truly seen the movements of the scene as when you have listened to the phonograph you have heard the voice which once impressed the record of its vibrations. Of all the animatograph pictures those that appeal most to me are the natural scenes, such as the waves rolling up into a sea-cave and breaking on the rocks at its mouth, and dashing foam and spray far up into its interior. Nothing is wanting to complete the illusion, save the reverberating roar of the waves.

Note.—Since the delivery of these lectures, Mr. Shelford Bidwell, F.R.S., has pursued the subject of the curious colour - effects mentioned at the foot of p. 96, and has reached some extraordinary results. A cardboard disk 8 inches in diameter is half-covered with black velvet, the other half being left white or gray. A sector of 45° is cut away at the junction of the black and white portions, and this disk, suitably balanced, is mounted upon a revolving apparatus to rotate with a speed of about 6 to 8 turns per second. Behind it, so as to be visible at each turn through the gap where the sector has been cut away, is placed a coloured picture, and a bright lamp is placed a few inches in front to illuminate it. The direction of rotation is such that the open sector is preceded by black and followed by white. On thus viewing a picture by intermittent vision, each part appears of a pale colour complementary to its actual tint. A red rose with green leaves appears a green rose with reddish leaves. A blue star on a yellow ground appears as a yellow star on a bluish ground. Black printing on a white paper appears whitish printing on a grayish paper, and so forth.

APPENDIX TO LECTURE II

ANOMALOUS REFRACTION AND DISPERSION

ON p. 78 attention is drawn to the circumstance that the spectrum as produced by a prism is irrational; that is to say, that the dispersion is such that the different waves are not spaced out in proportion to their wave-length, the red and orange waves being relatively crowded together at one end of the spectrum, while the violet and blue waves are unduly spread out. But the dispersion is different on different substances. In fact, no two substances disperse the light in exactly the same way, though in general the order of the colours is the same, and the general trend of the irrationality is to compress the red end. But there are a few known substances in which this irrationality becomes excessive, and develops into an entirely abnormal dispersion in which the violet waves are less refracted than the red! This phenomenon of anomalous dispersion was first noticed [1] in 1840 by Fox Talbot in some crystals of the double oxalate of chromium and potassium. The colours of the spectra of some of these crystals were so anomalous that he could only explain them "by the supposition that the spectrum, after proceeding for a certain distance, stopped short and returned upon itself." In 1861 Le Roux found that vapour of iodine, which transmits only red and blue, actually retards the red more than the blue, and gives an inverted spectrum. Christiansen in 1870 noticed that an alcoholic solution of magenta (rosaniline) has an ordinary refraction for the waves from red to yellow, the yellow being

[1] See Tait's *Light*, p. 156.

refracted more than orange, and orange than red, but it absorbs green powerfully, and all the rest of the colours —commonly called more refrangible—are in this substance refracted less than the red! In this case, the spectrum literally returns back upon itself. Other observations have been added by Kundt and others. In particular, Kundt discovered that some of the metals, when made up into excessively thin prisms, possess an anomalous dispersion.

The first point to note in discussing this phenomenon of anomalous dispersion is that it only occurs in highly coloured substances. It is closely related to the circumstance that in these substances there is, by reason of their molecular constitution, a strong absorption for waves of some particular wave-length. Thus in rosaniline, which has a strong absorption-band in the green, the fact that green light is absorbed appears to exercise a perturbing influence upon the waves of the shorter kinds, causing them to be less refracted instead of more. These remarkable phenomena obviously have something to do with the way in which the molecules of ponderable matter are connected with the ether. None of the dispersion formulæ of Cauchy, Ketteler or others gave a satisfactory account of them.

In 1872 Lord Rayleigh considered the problem of the refraction of light by opaque bodies, and in the *Philosophical Magazine* (vol. xliii. p. 322) gave the following exceedingly suggestive comment :—

"On either side of an absorption-band there is an abnormal change in the refrangibility (as determined by prismatic deviation) of such a kind that the refraction is *increased* below (that is on the red side of) the band, and *diminished* above it. An analogy may be traced here with the repulsion between two periods which frequently occurs in vibrating systems. The effect of a pendulum, suspended from a body subject to horizontal vibration, is to increase or diminish the virtual inertia of the mass according as the natural period of the pendulum is shorter or longer than that of its point of suspension. This may be expressed by saying that if the point of support tends to vibrate more rapidly than the pendulum it is made to go faster still, and *vice versâ*. Below the absorption-band the material vibration is naturally the higher, and hence the effect of the associated matter is to increase (abnormally) the virtual inertia of the ether and therefore the refrangibility. On the other side the effect is the reverse."

In 1893 von Helmholtz published a remarkable study [1] based on Maxwell's electromagnetic theory of light. The essence of this theory is as follows :—

The electromagnetic waves passing through the ether travel at a rate which is retarded by the presence of material molecules, the ether being as it were loaded by them. These heavy particles cannot be set into vibration without taking up energy from the advancing wave ; and so long as there is no absorption, they give up this kinetic energy again to the wave as it passes on. In this way the velocity of propagation of the train of waves is slightly less than the velocity of propagation of the individual wave ; the front wave of the train continually dying out in giving its energy to the material particles in the medium. In such a medium there will of course be ordinary refraction ; and as the velocity of propagation of the wave-train will depend on the frequency (*i.e.* on the wave-length) of the oscillations (there being in general a greater retardation of the waves of higher frequency, *i.e.* of shorter wave-length), there will be a dispersion of the ordinary kind. All this applies to waves the wave-length of which is large compared with the size of the molecules. But if there were smaller waves, the frequency of which coincided very nearly with the natural oscillation-period of the molecules or atoms, such waves would set up a violent sympathetic vibration of these material particles, and would be strongly absorbed. Suppose that there are waves of still smaller size and still higher frequency. Their oscillations are too rapid to affect the atoms ; they pass freely between the interstices of

---

[1] See *Wiedemann's Annalen*, xlviii. p. 389. The fullest account of this that has appeared in English is in *The Electrician*, xxxvii. p. 404, and an abstract account by Professor Oliver Lodge, *ib.* p. 371 (July 1896).

matter and are not retarded, therefore not refracted or dispersed, or only very slightly so. The medium would act as if almost perfectly transparent to such waves ; and their refraction might be either slightly negative or slightly positive ; whilst for the minutest waves of all the refraction would be simply zero. The formula which von Helmholtz deduced is in its simplest form the following :—

$$\mu^2 = \frac{a^2 - n^2}{\beta^2 - n^2},$$

where μ is the refractive index of the medium (supposed quite transparent), n the frequency, and a and β constants depending on the material. To interpret the formula, consider what it reduces to in the following cases (1) n much smaller than a or β; (2) $n = \beta$; (3) $n = a$; and (4) n much greater than a or β. In the first case, n being small we are dealing with long, slow-period waves such as Hertzian waves or those of infra-red dimensions. Neglecting n^2 compared with a^2 or β^2 the formula reduces to $\mu = a/\beta$, being independent of wave-length. In the second case if the frequency is such that $n = \beta$ the medium cannot possibly be transparent, as there would be violet absorption. The real meaning is that as n increases from case (1) toward the value $n = \beta$ the refractive index increases, and would become indefinitely great were it not for the absorption that sets in. In the state of things between cases (2) and (3), where n is larger than β but smaller than a, the values of μ calculated by the formula are imaginary ; but owing to the absorption they would in reality diminish down to near zero, that is to say, the refraction in these conditions becomes negative. This corresponds to the state of things observed by Kundt with their refracting prisms of iron, nickel, and platinum, which refract the light *toward* the refracting edge instead of *from* it. As n increases from case (3) when it equals a, the zero value of μ gradually changes, and, when n becomes very great compared with a and β, it approaches to unity, so that for excessively short waves there is no refraction at all.

Consider the particular value of n for which μ becomes a maximum. This is the case in which the excessive absorption makes the medium practically opaque. For values of n a little less than this there will be practically complete transparency and ordinary refraction and dispersion; for values of n a little greater than this there will again be practical transparency, but there will be a refraction in the wrong direction (μ being less than unity), and the dispersion will be anomalous. Fig. 63 illustrates this dependence of μ upon n. In the case of rosaniline, the frequency for which the absorption becomes excessive is about 578 billions per second, corresponding to a wave-length

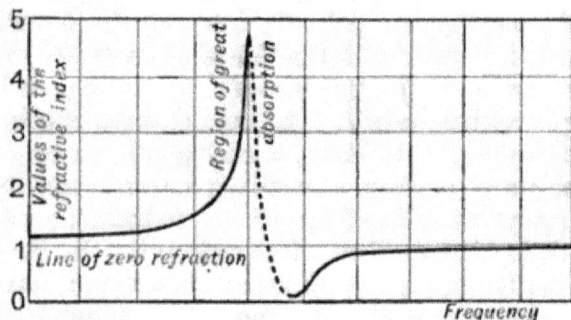

FIG. 63.

of 21 millionths of an inch in air. E. F. Nichols has found that quartz shows a similar change of properties for infra-red waves of a frequency between 36 and 45·4 billions per second. For almost all ordinary transparent substances the absorption-band occurs a long way down in the ultra-violet; in some it may possibly occur in the infra-red. It may be possible, for example, for flint-glass, the refractive index of which for ordinary light is between 1·5 and 1·7, to have a refractive index as high as 2·6 for disturbances of very low frequency such as Hertzian waves: that being the theoretical value for long waves as required by Maxwell's theory to correspond to the square-root of the observed dielectric constant. Probably many substances have more than one absorption-band, thus still further complicating the anomalous dispersion.

LECTURE III

POLARISATION OF LIGHT

Meaning of *polarisation*—How to polarise waves of light—Illustrative models—Polarisers made of glass, of calc-spar, and of slices of tourmaline—How any polariser will cut off polarised light—Properties of crystals—Use of polarised light to detect false gems—Rubies, sapphires, and amethysts—Polarisation by double refraction—Curious coloured effects, in polarised light, produced by colourless slices of thin crystals when placed between polariser and analyser—Further study of complementary and supplementary tints—Exhibition of slides by polarised light—Effects produced on glass by compression, and by heating.

SCIENTIFIC men often fall into the habit of using long and difficult words to express very simple and easy ideas. The natural consequence is, that people are often led to think that there is something difficult about a really easy subject, whereas the main difficulty is to understand the meaning of the words selected to describe it.

The word "polarisation," used in optics, is one of these terms. It sounds very learned and difficult, but the idea it is intended to express is really very simple. Let me try to give you the idea before we try to fit any name to it.

In my first lecture I endeavoured to give you some simple notions about waves and the way they travel. I asked you particularly to distinguish between the oscillatory motions of the particles and the forward travelling of the waves themselves. Let us return to the motions of the particles. Suppose any particle or group of particles to have motion given to it, a rapid "to-and-fro"—in other words, let it be supposed to vibrate. Then if it is surrounded by a suitable medium, and its vibrations occur with a sufficiently great frequency, it will set up waves in the surrounding medium which will start off from it, travelling away at a definite speed depending on the rigidity and density of the medium. In the case of a compressible surrounding medium such as air, the vibrating body (if vibrating between the limits of frequency appropriate for sound—that is to say, between about 30 per second and 38,000 per second) will compress the air in front of itself as it moves forward, and rarefy the air behind it as it moves back, with the result that it sends off waves of condensation and rarefaction. If, as in Fig. 64, the oscillating body is a sphere moving rapidly up and down along the short path AB, it will tend alternately to condense and rarefy the air above and below it, and these compressions and rarefactions will travel off upwards and downwards, spreading a little as they go. But hardly any waves of compression and rarefaction will travel off sideways from the oscillating sphere, because in oscillating up-and-down it does not either condense or rarefy the air at its sides. The wave in this case would be described as a *longi-*

FIG. 64.

tudinal wave, meaning one which is propagated along the line in which the particular motion exists—in this case vertical. If you want to know more about the travelling of sound-waves, you must read Professor Tyndall's delightful book *On Sound ;* or if you are deep students you will study Lord Rayleigh's two mathematical volumes on the *Theory of Sound.*

But now, suppose that you have to deal not with a medium like air that is compressible, but with a medium like jelly that is incompressible, and in which the density is small compared with the rigidity that it opposes to any rapid shear. If in this case you set up an oscillation with a sufficiently great frequency, waves will be set up which will travel off at a high rate, but not in the line of the

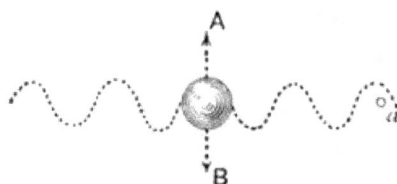

Fig. 65.

motion. On the contrary, they will travel off sideways in all directions in ripples. Let Fig. 65 represent a sphere embedded in the middle of a great block of surrounding jelly, and that it is made to oscillate up-and-down as before, but with a great rapidity. It cannot move up without tending to tear or shear the jelly all around its girth, nor can it move down without tending to tear or shear the jelly downwards; and these shearing stresses travel outward in all directions, so that a particle at *a* will, as these waves or ripples in the solid jelly reach it, tend also to move up and down. In any medium, whether a jelly or not, if the particles are in such relation to one another that the movement of any of them tends to set up a shearing stress, then that medium will, like the jelly, propagate the disturbances sideways. The waves in such cases would be described as *transverse* waves — meaning waves which are propagated in directions at right angles to the direction in which the to-and-fro displacements are executed.

Now the waves of light are of the transverse kind; and though they can pass through air, are not waves of the air as sound-waves are. Waves of light can cross the most perfect vacuum; they travel thousands of millions of miles in the vacuous space between the stars. They are waves of another medium which, so far as we know, exists all through space, and which we call, using Sir Isaac Newton's term, *the ether.* If you ask me what the ether is made of, let me frankly say I do not know. But if light consists of waves, and if those waves can travel across the millions of miles that separate the stars from the earth, then it is clear that they must be waves of *something;* they are not air-waves nor water-waves, because interstellar space is devoid both of air and of water. They are waves of a medium which, though millions of times less dense than water or air, has yet a property that resists being torn or sheared asunder; exceeding the resistance to shear even of hard-tempered steel. Though it is not a jelly, since things can move through it more freely than you or I can move through the air, yet it resembles the jelly in this property of resistance to shear, and propagates vibrations transversely to the direction of their displacements. Though we know neither the density of the ether (though it must be very small) nor its rigidity to shear (which must be very great), we do know something which depends on the ratio of these two properties, namely, the velocity of propagation of those ether-waves which we call " light " (see Appendix, p. 156).

Well, now having got this notion about transverse waves, let us go back to the wave-motion model which we used in the first lecture. It has, as you will remember (Fig. 2, p. 8), a row of little white particles, along which row the wave is propagated from left to right, though each little particle moves up and down. It is, therefore, a model of a transverse wave; the direction of travel of the wave is at right angles to the direction of the displacements.

But if a wave is to travel along a line of march, from A to B, we may fulfil the condition of transverse vibration in other ways. The small to-and-fro motions must be executed *across* the line of march; and they may be across the line of march without being vertical—they may be horizontal, or oblique. If I turn my wave-motion model on its side, the little white particles now move horizontally toward you and from you, but the wave still travels from left to right.

If I stretch across the room a long indiarubber cord, holding one end of it in my hand, I can throw it into transverse vibrations. If I move my hand rapidly up-and-down, I produce up-and-down vibrations. If I move my hand right-and-left, I get right-and-left vibrations. If I move my hand obliquely to-and-fro, I produce oblique vibrations; and the cord transmits them all.

Now, all that the word *polarisation* means is that the

motions are being executed in some particular transverse direction. If the vibrations are polarised vertically, that means that they are up-and-down waves that are travelling along. If I say that the light is polarised horizontally, all I mean is that the motions are executed

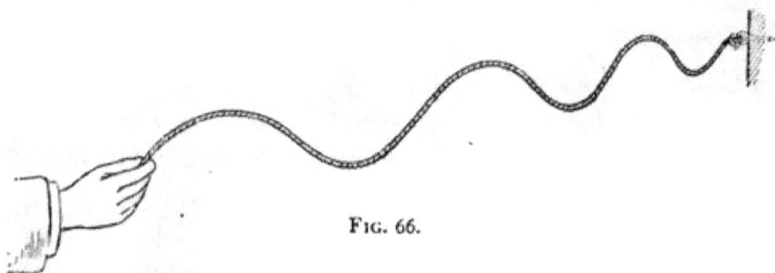

FIG. 66.

right-and-left across the line of march. Can anything be simpler?

Here is a lump of jelly. It will serve excellently to show how a polarised vibration is propagated. I stick into it horizontally two pins with silvered heads—one at

FIG. 67.

one side, the other at the other. If I give a sudden displacement to one pin it quivers, and the jelly carries on the motion to the other. And note, if I strike one so as to make it quiver up-and-down, the other quivers up-and-down—here we have a vibration polarised vertically. If I make one quiver right-and-left the other quivers right-and-left—here we have a vibration polarised horizontally. If I make one quiver circularly round and round, the other quivers round and round also; giving an illustration of a circularly polarised vibration.

Now let us go to the waves of light themselves. If you look at a beam of white light you cannot by the eye [1] tell whether it is polarised to move up-and-down, or right-and-left. In fact you cannot tell whether it is polarised at all. Naturally, if the waves are so excessively small, and vibrate so many millions of millions of times a second, your eye cannot catch their motions.

The fact is that light of any natural kind, whether from the sun, an electric lamp, a flame, or any other source, is non-polarised; that is to say, it consists of vibrations which are not specially directed either up-and-down or right-and-left, or in any other one direction. Natural light, given out by hot bodies, is absolutely miscellaneous. Not only does it consist, as we saw in the last lecture, of a lot of different colours—that is, of different wave-lengths, mixed up together—but it consists of waves whose direction of transverse vibrations are also all jumbled up. At one instant they may be up-and-down; then they change to right-and-left, or to oblique, or circular, or elliptical, or possibly to something still more complex. Just think how the light starts from the white-hot tip of the carbon pencil in my electric lamp. The particles of white-hot carbon are in fierce vibration, jostling against one another, and in jostling impart vibrations to the ether—setting up

---

[1] Not as thrown on the screen. But the eye can be trained to detect the plane of polarisation, for example, of light from the blue sky, which is naturally polarised in directions at right angles to the position of the sun. The training consists in being able to recognise certain appearances called "Haidinger's brushes" which result from the feeble polarising properties of the refracting structures of the eye.

ether-waves. When any one particle gets a sudden jolt it quivers, and gives out a vibration, which we may represent by the curve (Fig. 68), with a lot of little wavelets each like its fellow, perhaps several thousands[1] of them before they die away. Each such vibration would die away like the note of a piano-string struck and left to itself. But perhaps before the motion has died away another jolt sets it off vibrating in a new direction, again to die away. Suppose millions of these little particles, all jostling, and vibrating, and sending out trains of wavelets. It is clear that one ought to expect the utmost admixture of wave-sizes and directions of vibration in the resultant light.

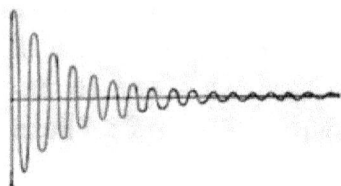

FIG. 68.

Then, you understand, that as natural light is not polarised in any particular direction, if we want to get polarised light we must do something to it to polarise it. But how?

[1] According to the researches of Fizeau, at least 50,000, on the average, in ordinary light. Prof. Michelson's more recent experiments, in which he has obtained interference between two waves the paths of which differed by more than 20 cm. or 1,000,000 wavelengths, prove that the average number of wavelets in each train must be reckoned in millions.

[TABLE

TABLE III.—POLARISERS

Principle.		Nature of Apparatus.	Reference.
By Reflexion . .	I.	Black glass at about 57°	(p. 153).
	II.	Delezenne's Polariser .	(p. 123).
By Refraction .	III.	Glass sheet at about 57°	(p. 154).
	IV.	Bundle of thin glass sheets set obliquely	(p. 154).
By Double Refraction	V.	Rhomb of Iceland Spar	(p. 120).
	VI.	Double-image Prism .	(p. 125).
By Double Refraction, with Absorption .	VII.	Slice of Tourmaline .	(p. 119).
By Double Refraction, with Internal Reflexion	VIII.	Nicol's Prism and its modern Varieties .	(p. 121).

In Table III. I have set down some eight different ways of polarising, which we will presently consider in their order. But before we deal with any of them, let us go back to the vibrations of cords and see how they can be polarised.

Here (Fig. 69) is an indiarubber cord passing through a wooden box with vertical partitions. These partitions limit the movements and only allow vertical vibrations to pass through. If I vibrate the cord in any way, it is only the vertical components of the vibration that succeed in getting through. The waves, after passing through the box, come out polarised in a vertical plane. If I turn the box over on its side (Fig. 70) it will now transmit only horizontal components of vibration. What will happen, then, if I pass the cord through a second box, as in Fig. 70? That depends on the positions of the boxes. If the first one P is set with its partitions

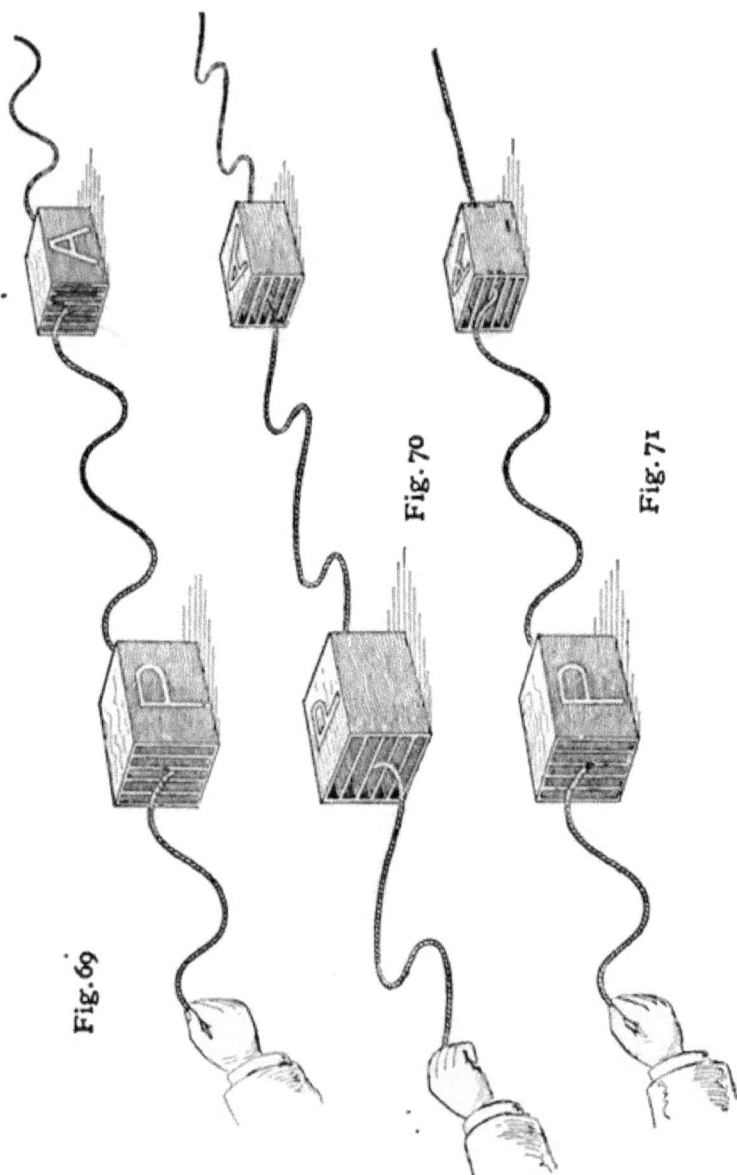

Fig. 69

Fig. 70

Fig. 71

vertical, it will polarise the waves vertically, and as these waves travel on they will come to the second box marked

A.   If this also has its partitions vertical, the vertical
waves will get through it also.   If both boxes are turned
over on their side, then the first one will polarise the
waves horizontally, and the horizontally polarised waves
will pass through both boxes.   But if I have the first box
P set vertically and the second box A horizontally (Fig.
71), P will polarise the vibrations so that they will not
get through A, but will be cut off.   However P is
placed it will polarise the waves; if A is turned so as
to cross the waves they will be cut off.

Upon the lecture table is another model which illustrates the same set of facts more fully.   If you understand it you will have no difficulty in understanding the
optical apparatus that we are going to use.   In this
apparatus the vibrations of a thin silk cord—best seen
by those in front of the table—are produced by attach-
ing one end to the prong of a tuning-fork, the vibrations
of which are maintained by an electromagnetic attach-
ment.   To the distant end of the cord is attached a small
weight, which has been so adjusted that the cord is thrown
into stationary waves.   In brief, the vibrations of the cord
are tuned to those of the fork.   To polarise the vibrations,
the motions of the cord are confined by means of a pair
of glass plates mounted in wooden cylinders (Figs. 72,
73).   At the first nodal point of the cord the first pair
of glass plates acts as a polariser, P ; the cord beyond
that point vibrating in the plane thus imposed upon it.
A pointer fixed upon the wooden cylinder shows the
direction of the plane of polarisation.[1]   The second

―――――――

[1] Concerning the term, "plane of polarisation," see remarks in
Appendix to this Lecture, p. 158.

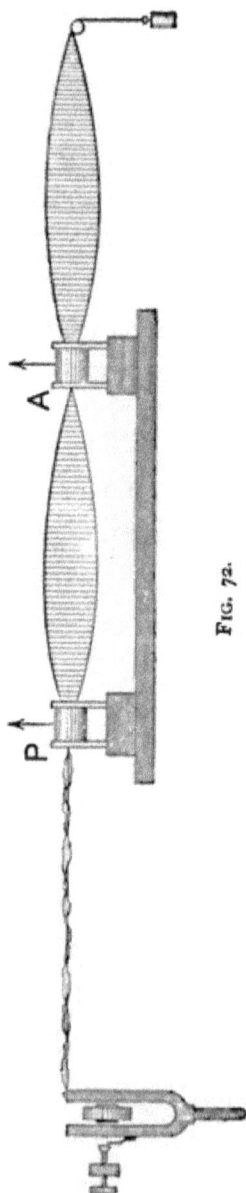

FIG. 72.

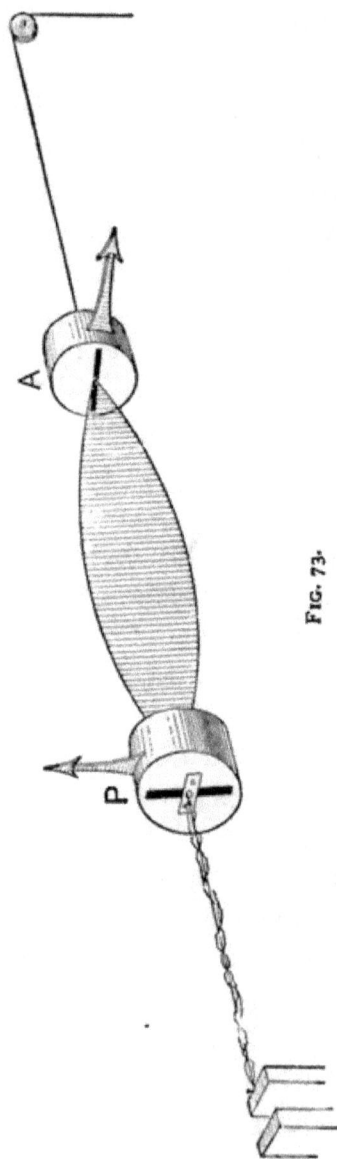

FIG. 73.

pair of glass plates is set at the second nodal point
to act as an *analyser*, A. The vibrations of the cord

are made vertical by the polariser P, and when the
plane of the analyser A is also vertical (as in Fig. 72)
the vibrations which pass through the polariser pass
through the analyser also.   But, if (as in the previous
experiment with the boxes) the analyser is turned round
a quarter, so that the slit between the glass plates lies
across the vibrations (as in Fig. 73) the vibrations are
no longer transmitted.   To recapitulate, *the vibrations
are transmitted when the polariser and analyser are
parallel to one another: but are cut off and extinguished
when polariser and analyser are crossed.*   Hence, by
turning round the analyser to such a position that it
cuts off the vibrations we can ascertain with accuracy[1]
the direction of the vibrations proceeding from the
polariser.

But why should we linger longer upon mere models
when we can operate with light-waves themselves?
My assistant throws upon the screen a beam of white
light from the electric lamp within the optical lantern.
He now places in the path of the beam a large polariser,
P (Fig. 74).   What this polariser is, I will presently
explain.   He now sets it so that it polarises the light,
allowing to fall upon the screen those waves only whose
vibrations are executed in a vertical plane.   The white
disk of light on the screen consists, in fact, of up-and-
down light only.   Your eye would not tell you whether
the light was vibrating up and down, or even that it was

---

[1] The model will enable the orientation of the plane of the vibrations to be determined to within about half a degree of angle.   That
is, if the analyser is as much as half a degree out of the crossed
position, the vibrations are not completely extinguished.

polarised at all.   To ascertain that the waves are really
polarised we must have recourse to an analyser.   This
analyser, A, is itself simply a smaller polariser.   In
order that you may see it the better it is mounted
(see Fig. 75) by thin strings upon a ring-support,
the shadow of which you see on the screen.   If this
is also set in the proper position to transmit up-and-
down vibrations, the polarised light will come through

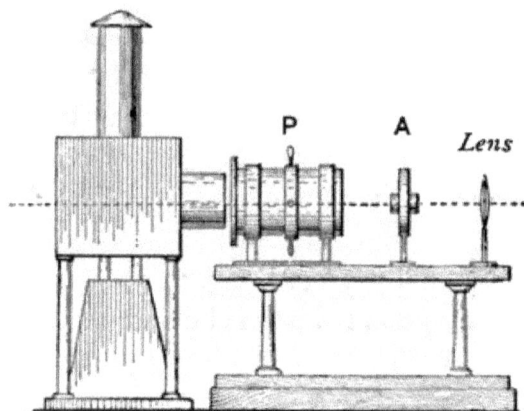

FIG. 74.

it, both polariser and analyser being clear as glass.   If
now the analyser A is turned round one quarter it will,
though clear as glass, entirely cut off the up-and-down
vibrations, with the result (Fig. 76) that no light gets
through it.   This cutting off of the light by turning
the analyser one quarter round *proves* that the light was
polarised.   When the planes of polariser and analyser
are parallel to one another—both vertical, or both
horizontal,—then we have the "bright field" of transmitted light.   When the planes of polariser and analyser

are crossed—one vertical, the other horizontal—then
the light is cut off, and we have the "dark field."

There is a gem called the tourmaline which, when
cut into thin slices, has the property of polarising light.
This gem[1] is often found of a dark green colour, but
also of brown, dark blue, and even ruby tint. Into the
beam of ordinary white light now cast upon the screen

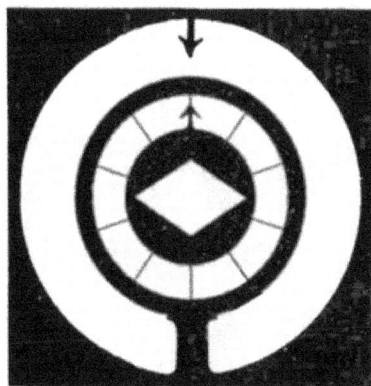
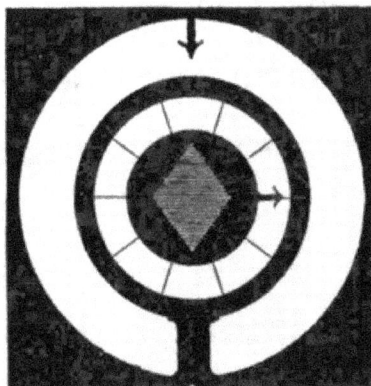

FIG. 75.   FIG. 76.

there is now introduced a thin slice of brown tourmaline
(Fig. 77). It looks dark, for it cuts off more than half
the light. But such light as succeeds in getting through
is polarised—the vibrations being parallel to the longer
dimension of the slice. A second thin slice of tourmaline
is now introduced, and superposed over the first. When
they are parallel to one another light comes through
both of them (Fig. 78). But if one of them is now

[1] The dark green tourmaline is also sometimes called the Brazilian
emerald, though it is of entirely different composition from an
emerald. The bishops of the South American Catholic churches
wear tourmalines in their episcopal rings, instead of emeralds.

turned round, so that they are crossed, as in Fig. 79, no
light can get through the crossed crystals. The one cuts
off all horizontal vibrations and horizontal components
of vibration, the other cuts off all vertical vibrations and
vertical components of vibration. Hence, when crossed,
they produce a "dark field." One acts as polariser, the
other as analyser.

Let us return to the big polariser (Fig. 74) which we
used in the previous experiment, and which was as clear
as glass. It is made of Iceland spar, a natural crystal,

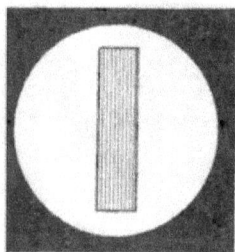 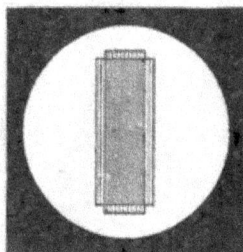 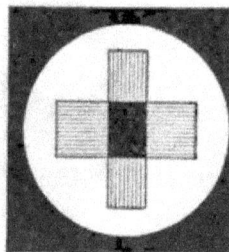

Fig. 77.   Fig. 78.   Fig. 79.

which once was common but now is rare and expensive.
As imported from the mine in Iceland this spar possesses
the peculiar property known as "double refraction":
when you look through it you see everything double.
Here is a fine specimen mounted in a tube. Look at
your finger through it; you will see two fingers. It is a
substance which splits the waves of light into two parts,
giving two images; and, moreover, polarises the light in
the act of splitting it, so that each part is polarised.
We do not, however, want both images; we want only
one. What do we do? We adopt the method proposed
eighty years ago by William Nicol, a celebrated Scotch

philosopher, and construct out of a crystal of the spar
a "polarising prism," or Nicol prism.   Here are several

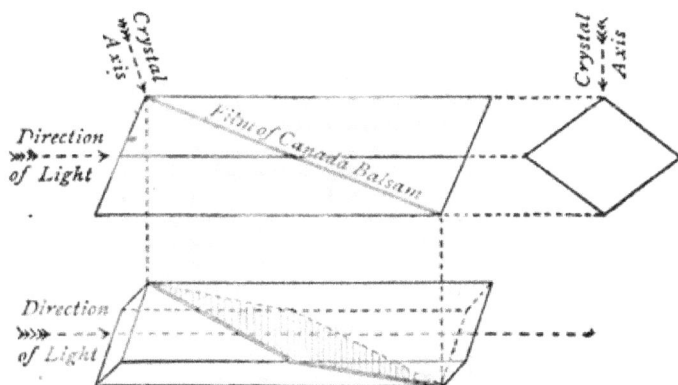

F.G. 80.

Nicol prisms of various sizes ; and also several modern
modifications [1] of the Nicol prism.   Here also is a large
wooden model to illustrate Nicol's method.

[1] In Foucault's modification, a film of air is interposed between
the two wedges of crystal.   In Hartnack's prism a film of linseed
oil is interposed, and the ends of the wedges are squared off.   I have
myself from time to time suggested several modifications which are

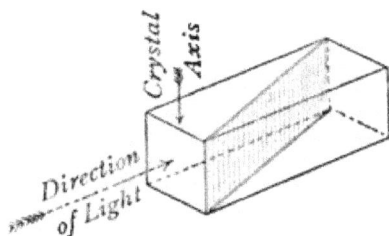

FIG. 81.

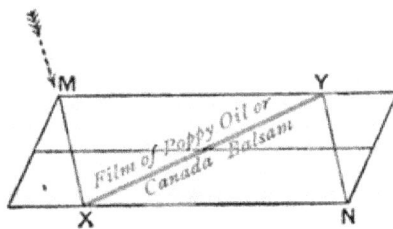

FIG. 82.

improvements upon the original Nicol prism.   In one of these, the
natural end-faces of the prism are sliced away parallel to the
crystallographic axis so as to leave terminal faces that are " principal
planes " (Fig. 81), and the crystal is then sliced with an oblique cut

Selecting a piece of Iceland spar of suitable proportions we slice it across (with a piece of copper wire,
used as a saw, and some emery powder) in an oblique
direction from one of its two blunt corners to the other;
polish the surfaces, thus dividing the prism into two
wedges.   These are then cemented together again
with Canada balsam (a resinous cement); and the
polarising prism is complete.   Its operation upon light
is as follows.   When the waves enter through one end-
face they are split into two parts which take slightly
different directions, and strike at different angles upon
the film of balsam.   As a consequence one of the two
beams when it meets the film of balsam is reflected
off sideways, as from an oblique mirror, while the
other goes through the prism and emerges at the other
end-face.   Consequently only one of the two beams
gets through the prism, the other being suppressed or
reflected out of the way.   Prisms made in Nicol's way

that is also a principal plane, and these wedges are then reunited

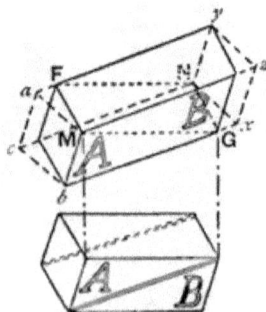

Fig. 83.

with Canada balsam or linseed oil. In a
cheaper modification—a "reversed Nicol"
—the natural end-faces are cut off (Fig.
82) so as to reverse the shape, and the
oblique cut is then made along a reversed diagonal and is nearly in a
"principal plane." In a third modification the end-faces are first trimmed off
obliquely as principal planes of section
through one of the natural edges of the
end-face; an oblique cut is then given
(as in Fig. 83) between two of the
terminal arrêtes, from FM to GN, and the two pieces are then
transposed; and they are finally reunited by balsam along two of
their natural faces.

have usually oblique end-faces of diamond shape. The vibrations which pass through are those executed in the direction parallel to the shorter diagonal (Fig. 84); while those which are suppressed are those parallel to the longer diagonal. The large polariser used in front of the lantern (Fig. 74, p. 118) is simply a large Nicol prism.[1]

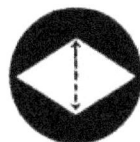

FIG. 84.

[1] In consequence of the dearth of spar, large Nicol prisms can only be procured at extravagant prices. In 1888 Mr. Ahrens constructed for me a large reflecting polariser, having a clear aperture of $2\frac{3}{4}$ inches. For projection purposes it is quite equal to a Nicol prism of equal aperture, and is much less costly. In this reflecting polariser, which is constructed on a principle suggested by Delezenne, the light is first turned to the proper polarising angle (about 57°) by a large total-reflexion prism of glass cut to a special shape. It is then reflected back parallel to its original path by impinging upon a mirror of black glass covered by a single sheet of the thinnest patent plate glass to increase the intensity of the light. Fig. 85 shows the design of this prism. Compared with a large Nicol prism it has one disadvantage : it cannot be conveniently rotated, so that it polarises the light in a fixed plane. To obviate this defect, I devised an "optical rotator" to place on the end of the prism. This consists simply of two plates, q_1 and q_2, of "quarter-wave" mica ; the first of them being mounted with its axis fixed at 45° to the plane of polarisation ; the second q_2 being mounted in a revolving frame which can be turned to any desired position. The effect of rotating this plate is to rotate the plane of polarisation.

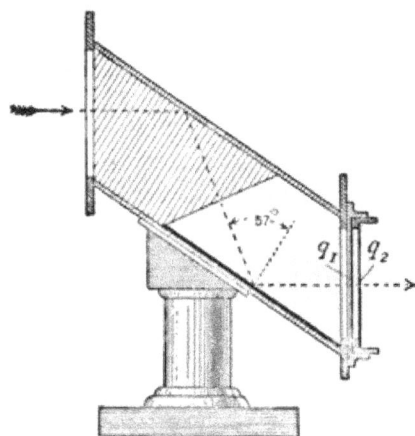

FIG. 85.

Let us devote a little further attention to this phenomenon of the double-refraction that thus yields us two
beams of light that are polarised in different planes.
Here is another wave-motion model (Fig. 86) constructed
to show two sets of waves which are polarised in planes
mutually at right angles to one another.   Here are two

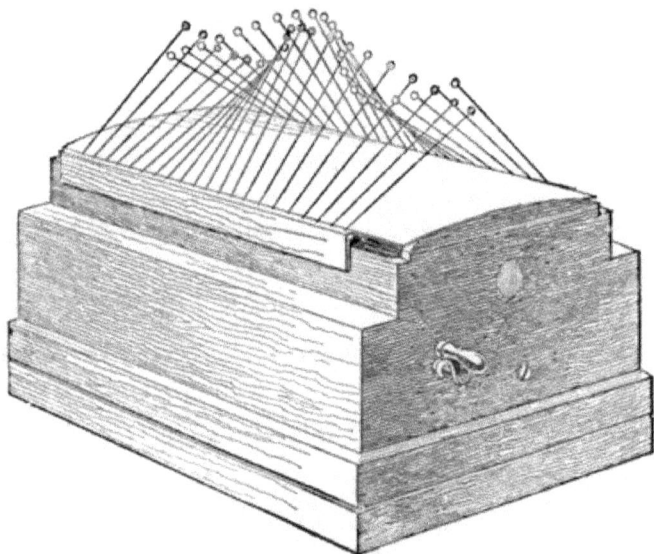

Fig. 86.

waves of silvered beads, both of which, when I turn the
handle of the model, will march along.   Both have the
same wave-length, both march at the same pace and
toward the same end.   But there is this difference
between them : in one the displacements are polarised
at 45° one way ; in the other the displacements are at
45° the other way.   Of course there is some mechanism
inside the box to make them move thus ; but they illus-

trate what is meant by saying that there can be two waves polarised at right angles to one another.

But the point still remains why should the spar so act on the light as to split it into two oppositely polarised beams? Let us first prove that the spar really does this. Here is a piece of spar mounted as a double-image prism.[1] In front of the lantern is placed first a metal diaphragm having a round hole in it, the image of which is focused on the screen, giving a circular white spot. Now interposing the double-image prism we see that it splits the light into two parts, diverting half the light away from the original spot, and producing a second one which (with ·this size of aperture in the diaphragm) overlaps the first. On rotating the double-image prism it is seen that the ordinary image remains stationary, whilst the extraordinary image revolves around it. Now to prove that they are each polarised, I interpose a Nicol prism. On rotating it, it is observed to cut off first one of the two spots and then the other. If the Nicol prism is set so as to

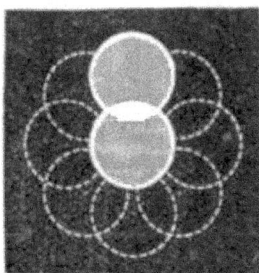

FIG. 87.

---

[1] That is a prism of spar mounted in such a way as to throw both the images upon the screen. There are several modes in which such may be constructed. The usual mode is to take a wedge of spar cemented to a corresponding wedge of glass. The "extraordinary" beam (that is to say, the beam whose vibrations are executed in planes parallel to the crystallographic axis) goes straight through the "ordinary" beam, and emerges obliquely, giving a displaced image. If the prism is made of two wedges of spar cut at different angles and crossed, the "ordinary" image is the central one while the "extraordinary" image revolves.

transmit only vertical vibrations, while the double-image prism is rotated it is observed that when, as in Fig. 88, the two images are in the position vertically above one another, the ordinary is cut off, while the extraordinary is transmitted. On turning round the double-image prism the ordinary image gradually appears, while the extraordinary fades away, until when the prism has been turned one quarter round the ordinary image is transmitted, and the extraordinary cut off, as is evident from Fig. 89. If the prism is turned to 45° both images

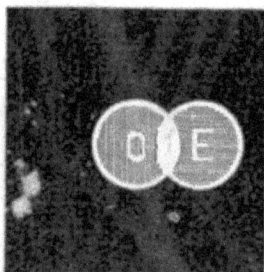
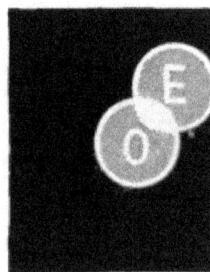

Fig. 88.   Fig. 89.   Fig. 90.

are equally bright, the directions of the resolved vibrations being as shown by the fine lines in Fig. 90.

Now, having proved that the spar does split the light into two beams in which the vibrations are moving in different ways, we have yet to consider how it effects this. The resolution of oblique movements into two components at right-angles to one another is an important principle in mechanics, and one which is best illustrated for our purpose by means of a model. Suppose that there is a displacement taking place obliquely, it can always be resolved into two parts—a part which is up-and-down, and a part which is right-and-left in

direction.   The model (Fig. 91) is fixed upon a board,
on the corner of which is drawn a little diagram.   Suppose the oblique motion to be from A to B, then you can
resolve that oblique motion into two parts—a vertical
part marked AV, and a horizontal part marked AH.
Every schoolboy knows the problem called the "paral-
lelogram of forces," according to which when two forces
act on one point at the same time, they combine into a

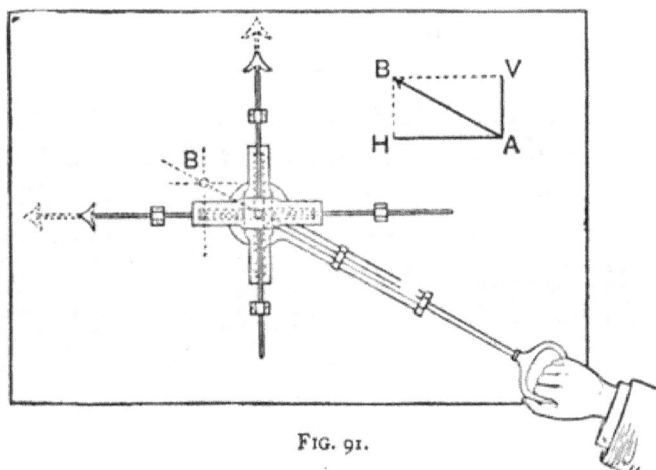

FIG. 91.

single oblique resultant along the diagonal of the paral-
lelogram of which the two component forces are sides.
Well our present problem is simply the converse to that.
Here in the model are two wooden slides, with cross-
heads.   One can slide up-and-down only : the other
right - and - left only.   Running in grooves in the two
cross-heads is a roller fixed to the end of a third bar of
wood, which can be set in any oblique position, and
then slid along obliquely by hand.   If I set this third
bar horizontal and slide it along, all the movement it

gives is horizontal—there is no vertical component. If I set it vertical and slide it up and down, it simply moves the vertical slide up and down. But if I set it obliquely, and slide it along, then part of its motion produces a movement of the vertical slide, while the horizontal component of its motion produces a movement of the horizontal slide. How much of the motion will be vertical, and how much horizontal, will depend obviously on the angle at which the original motion is imparted. Now there is one particular angle at which the resolved portions would be equal to one another. What is that angle? If I set the bar which produces the displacement exactly midway, at 45°, that displacement will be resolved into two equal components. You will remember that at 45° the double-image prism yielded two images of equal brightness.

Such a resolution into two parts can be produced on oblique waves of light by any of the crystals here, such as Iceland spar, quartz, mica, or selenite, provided the crystal possesses a particular kind of structure. The condition is that the crystal shall possess a greater optical rigidity—a greater stiffness that is—in one direction than in another. You know what one means when one talks about the grain in a piece of wood—how much easier it is to split in one direction than in another. But this grain, which depends on the fibrous structure, manifests itself in other ways than by ease of cleavage. A piece of wood is harder to bend along the grain than across it. And so with some crystals: they possess an invisible structure, a grain so fine that we cannot see the fibres or lines of structure; but that grain manifests

itself in various ways.   There are differences in ease of
cleavage, and also differences in rigidity[1] in different
directions.   This particular crystal—Iceland spar—has
a greater rigidity in one direction than in another, and,
as a result, any wave of light passing obliquely into it
is split into two portions, one having vibrations parallel
to the axis of greatest rigidity, and another portion
having vibrations parallel to the axis of least rigidity ;
therefore at right angles to the former.   That is why it
splits the light into two parts, and why those parts are
each polarised.   Some crystals, namely, diamonds and
garnets, are equally rigid in all directions, and therefore

[1] The word rigidity is here preferred, though in many treatises
it would be spoken of as "elasticity."   To say that the crystal
possesses different "elasticities" in different directions—which
is quite correct—would convey to many people an erroneous idea,
because of the incorrect way in which the word "elastic" is often
used.   Elastic does not mean that a thing can be easily stretched.
The hardest hard steel is more perfectly elastic than a bit of indiarubber, but it certainly is not so stretchable.   In saying that it has
elasticity we mean that however little we may succeed in compress-
ing or stretching it, it returns back, when released, to its former
shape or size.   In scientific treatises that substance is regarded as
having the highest co-efficient of elasticity which requires the
greatest stress to produce a given deformation or strain.   When we
are dealing, as in the case of transverse displacements, with motions
tending to *shear* (see p. 107 above) the medium, the particular
elasticity to be considered is the elasticity which resists shearing,
and for this the term *rigidity* is entirely appropriate.   In all this
optical work we are of course dealing not with the mechanical
elasticity, but with the optical elasticity, that is to say, with the
elasticity of the ether within the substance.   It is this which in the
crystals in question is regarded as greater in one direction than
in another.   The greater the rigidity the higher is the velocity of
propagation of the wave whose displacements are in that direction.
See Appendix to Lecture III. p. 156.

these do not show any double refraction, nor do they polarise the light.

Here, again, is a model to illustrate the splitting of the waves. Two thin flexible strips of ebonite, O and E (Fig. 92), are inserted into saw-cuts in the end of a cylinder of wood. In the other end I can fix a third strip, A. Notice, if you please, that O is set with its edge vertical, and is capable of vibrating right and left; while E is set with its edge horizontal, and can vibrate up and down only. Holding the wooden cylinder in my left hand, I apply my right hand to give a vibratory

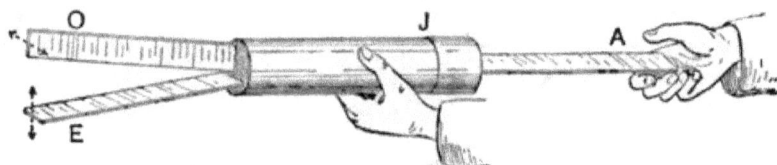

FIG. 92.

movement to the strip A at the other end. If I waggle A from right and left, then O vibrates from right and left. If I waggle A up and down, then E vibrates up and down, while O is quiet. If now I impart to A an oblique [1] motion, both O and E vibrate, the oblique motion being resolved into a vertical part and a horizontal part.

Returning to the waves of light themselves, let us now see some of the beautiful effects which result from these operations of splitting the vibrations into two parts, of recompounding them after passing through a

[1] The end of the cylinder into which A is fixed is capable of being turned round on a joint at J, the cylinder being made in two parts fitting on one another.

slice of crystal, and then of analysing—that is to say,
resolving—the light that falls upon the screen.

In front of the lantern there has been set the large
Nicol prism as polariser, and in front of that a smaller
Nicol prism as analyser. When the latter is turned to
cross the former we have the dark field (p. 119), all light
being cut off. The only light that passes through the
first Nicol consists of vertical vibrations, and as the
second Nicol is set to transmit horizontal vibrations only,
nothing comes through it. Now I take up a piece of
thin mica (that crystalline substance of which lamp-
shades are sometimes made, and often miscalled "talc"),
and hold it obliquely in the path of the polarised light
on its way from the polariser to the analyser—in fact,
between the two Nicols. See how it brings light into
the dark field. In Tyndall's expressive phrase it seems
to "scrape away the darkness." See, too, what beautiful tints it shows! Yet it itself is perfectly colourless.
Why, then, this light and these colours? Well, the
light is easily explained. The crystal possesses a
greater rigidity in one particular direction through
its substance — an "axis of maximum elasticity," as
it is called. If we set the slice with that axis
obliquely (Fig. 93) across the direction of the vertical
vibrations, then it will resolve those vertical vibrations
into two parts, one part parallel to the axis of maximum elasticity and the other at right angles to that
axis. These two sets of waves in the crystal will both
be oblique. They travel through the thickness of the
slice at unequal speeds, and when they emerge again at
the other side of the crystal one set of vibrations will

have got a little behind the other. Hence the two
components as they emerge and recombine will not
produce, as their resultant, vibrations that are vertical.
The resultant vibrations may be oblique, or even
elliptical; their precise nature and orientation will
depend on the nature and thickness of the slice of
crystal used, and upon the wave-length of the light.
But the immediate point is that the resultant emerging

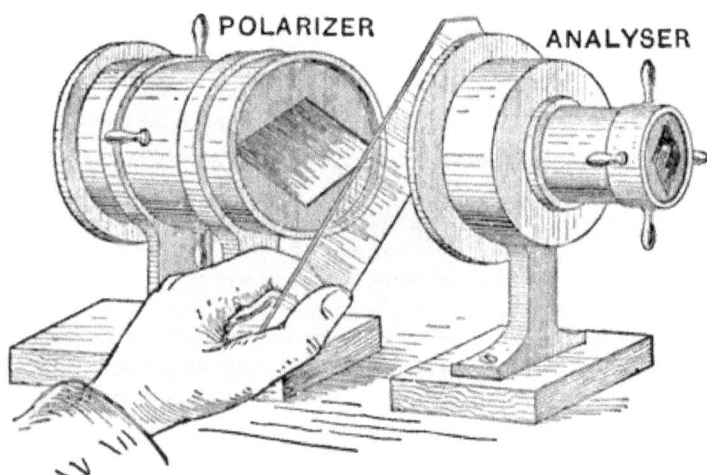

FIG. 93.

waves *will no longer be polarised vertically ;* the crystal
slice will have so split up, retarded, and recombined
the vibrations that they emerge vibrating in other directions. Hence this emerging light when it falls upon the
analyser will possess some horizontal components, and
these will, as you see, be transmitted. The colours
depend upon the thickness of the slice of crystal;
which in this case is very irregular, seeing that it is
merely a rough piece split off with a pocket knife.

This is mica; but there is another kind of thin crystal
known as selenite (or gypsum), which is also readily
split with the knife, and produces similar effects.   There
are plenty of other crystals that will produce similar
effects.

Here, mounted in a small glass cell, are two rubies.
Putting them in front of the polariser, and adjusting a
lens to focus them on the screen, you see how much
alike they are.   One is a genuine ruby, though slightly
flawed; the other is a sham ruby of glass, also slightly
flawed.   Which is which?   You may guess or choose;
but I wish to *know*.   That knowledge we can obtain
by putting in front of the lens the second Nicol as
analyser.   Turning it to position we obtain the dark
field; and in that field one of the two gems shines out
while the other is dark.   The one which shines out is
the genuine ruby, for it alone possesses the axis of
maximum elasticity.   If we slowly revolve the cell so
as to make the images of the two gems move around
one another, the one simply remains dark, while the
other goes through alternations of light and darkness.
It is dark in those positions in which its axis of maximum elasticity stands either vertical or horizontal, and
shines brightest in the two positions where that axis
stands at 45° of obliquity to right or to left, at which
angle the resolution into the two oblique components is
most complete.

Now we place in our apparatus, between polariser
and analyser, another cell containing several assorted
gems—a ruby, garnet, topaz, emerald, sapphire, chryso-
beryl, and a little diamond in the centre.   Adjusting

the analyser to give us the dark field, and then slowly
rotating the cell, note how each crystal, when its axis
of maximum elasticity comes to the oblique position of
45°, shines out. But there are two that show nothing—
the diamond and the garnet, crystals belonging to the
"cubic" system, whose elasticities in all directions are
equal.

Here are a few more objects for our polariscope,
slices of crystals and minerals. First of all a bit of
amethyst. Though by ordinary light it is quite clear
and of a pale purple tint, yet when examined by polarised light it is at once evident that the gem consists of
a number of superposed separate layers, which show
alternately dark purple and white; while some regions
of the crystal show strange masses of unexpected colour,
and these, if one turns the analyser round, change tint.
A second piece of amethyst, from Brazil, shows a more
perfect structure.

Here is a thin slice of gray Scotch granite. Its
natural mottlings are merely black and white, being
composed, as mineralogists will tell you, of small crystals
of transparent quartz and transparent felspar, mingled
with specks of black mica. But when viewed by
polarised light the whole slice shows wonderful gleams
of colour, and reveals new details of structure.

This next beautiful object is a thin transverse slice
of a stalactite, one of those natural deposits of calcare-
ous spar which hang like icicles from the roofs of caves.
Its deposit, layer by layer, almost concentrically, is
evident; but in the polariscope it shows a mysterious
black cross, which, when the analyser is rotated, changes

to a white cross. This black cross is visible in the dark field. Such black crosses one obtains whenever the object is one having different rigidities in the radial and tangential directions. The next slide shows it even better. This is an artificial crystal of a stuff called salicine, which can be dissolved in alcohol. When the solution is poured upon a warm piece of glass the alcohol evaporates, and the salicine crystallises. The

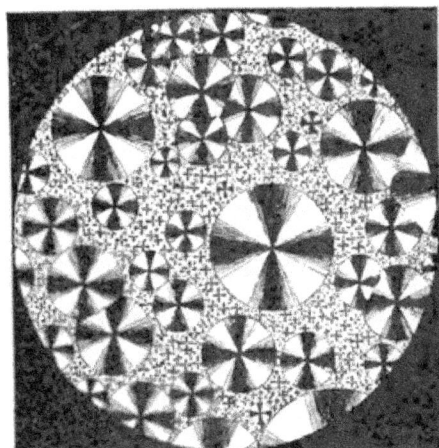

FIG. 94.

operation of crystallising starts at a number of independent centres, around which little groups of crystals grow with a radial structure and a circular outline. As their axes of maximum elasticity all point radially to the centre around which they grew, the maximum resolution of the wave light will occur in each little circle at 45° to right and to left, while in the two directions, vertical and horizontal, there will be no resolution of the polarised light, and these (in the dark field) will therefore remain dark, giving the black crosses. On revolving the analyser quickly, so that the black crosses change to white crosses and then turn black again in rapid succession, all the little crosses in the separate groups of crystals appear to revolve like so many little windmills.

Here is a thin slice of selenite, split off quite irregularly, and of different thicknesses in different parts. Without the analyser it shows on the screen nothing worthy of attention, being nearly clear as glass and quite colourless. But replacing the analyser to produce the dark field, at once a gorgeous set of tints is produced. At one part the thickness is such that the tint is a fine orange; just above it where the crystal is a shade thicker comes a patch of brilliant crimson. These being the tints in the dark field, see how, when the analyser is rotated a quarter so as to give the bright field, each tint changes to its complementary, the orange turning to azure blue, and the crimson turning to vivid green.

Now I have yet to explain how these colours come about. All I have said is that they depend upon the thickness of the crystal film, and upon the wave-lengths of the different kinds of light. If on its passage through the crystal the vibrations are split up into two parts that travel with unequal speeds, one set of vibrations will gain on the other; the one that is more retarded will, when it emerges, have lost step with the other set. This "difference of phase" due to the different speed of travelling may in a thin slice of crystal be as little as one quarter or one half only of a wave-length; or it

may, if the crystal is thicker, be more than a whole
wave-length, or it may be several wave-lengths. The
question as to how the two sets of waves will recombine
when they emerge at the other face of the crystal slice,
will depend upon the question how much one wave has
got out of step with the other wave. Clearly that will
depend on the kind of light and on the thickness of the
slice. If, for example, a slice of mica $\frac{1}{800}$ inch thick
caused the waves of yellow light to get exactly one
quarter of a wave-length out of step with one another,
then it is clear that a slice twice as thick would produce
twice as much retardation of phase, and make the two
sets of yellow waves get exactly half a wave-length out
of step. Further, if the thickness were such as to make
yellow waves get exactly half a wave out of step, it
would not produce an exact half-wave retardation upon
the larger red waves, or upon the shorter violet waves.
The consequence of all this is that in the recombina-
tions of the emerging waves after passing through any
given thickness of material, the angle at which the
vibrations recombine is different for different wave-
lengths. If, for example, the crystal thickness is such
that green waves in recombining come out vibrating
nearly vertically, then for that thickness of crystal green
light will be almost entirely cut off in the dark field, but
almost entirely transmitted in the bright field. So we
have this complementary relation between the tints in
the two positions of the analyser.

The succession of tints for a regularly increasing
thickness of crystal follows the order of the tints known
as Newton's "Colours of Thin Plates." Sir Isaac

Newton examined the colours of soap-bubbles and other thin films, and ascertained their relation to the

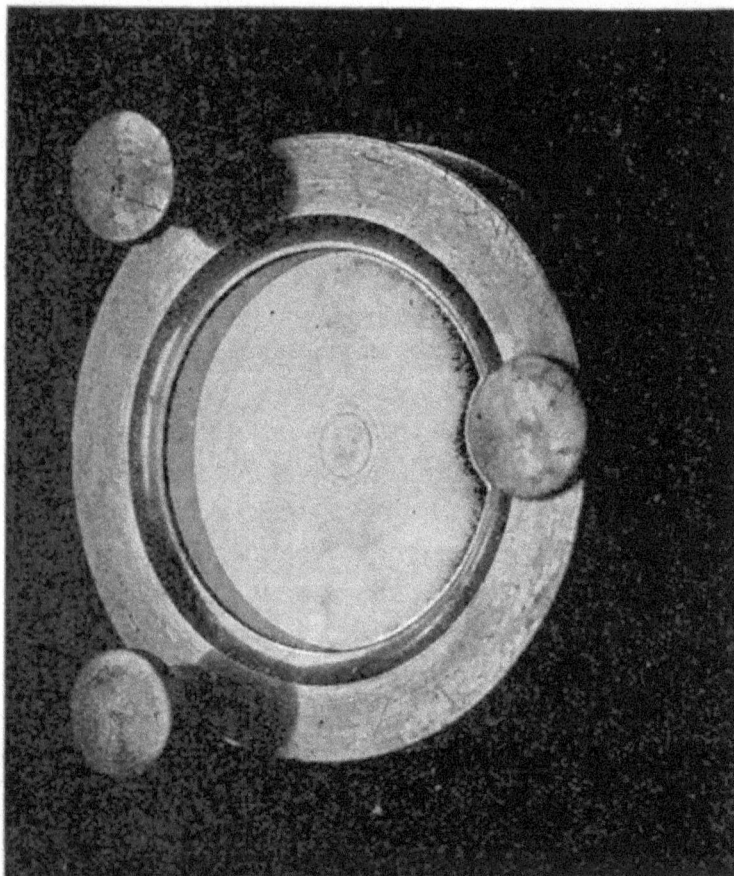

FIG. 95.

thickness of the film by ingeniously producing a film of air between two pieces of glass, one flat, the other very slightly curved. There is now projected upon the screen the system of coloured rings ("Newton's rings") so produced. Where the two pieces of glass touch there is a central dark spot; around this centre there are coloured rings of light of the several "orders." The tints are set forth in the accompanying Table. Those of the first order, beginning with black, are very dull, and end with a dark purple or "transition tint," after which the colours of the second order follow almost those of the spectrum, except that there is no good green. At the end of the second order comes again a purple "transition tint," after which the third order gives a sort of spectrum series with a good green but with a poor yellow. To the third order succeeds a fourth with paler tints, mostly green and red, then a still paler fifth, followed by higher orders that are still less distinct. The corresponding thicknesses of the film are given in millionth parts of an inch. For producing the kindred set of Newton's tints by means of thin films of selenite in the dark field of the polariscope much greater thicknesses must be used, because the phenomenon is due to the difference between the two velocities of the two sets of vibrations. Thus to produce a retardation of $\frac{1}{4}$ wave, and therefore give the same whitish tint as an air-film 5·5 millionths of an inch thick, a piece of selenite must be used the thickness of which is about $\frac{1}{1000}$ inch. The tints so produced—see p. 145—are not precisely identical with the air film tints because of the modifying effect of dispersion in the selenite.

[TABLE

TABLE IV

	TINTS OF NEWTON'S COLOURS OF THIN FILMS.	
Order.	Film Thickness.	Tint in Reflected Light.
I.	0	Black.
	3·5	Gray.
	5·5	Whitish.
	8	Straw.
	10	Orange.
	10·5	Brick Red.
	11	Dark Purple.
II.	11·5	Violet.
	13	Blue.
	15	Peacock.
	18	Yellow.
	19·5	Orange.
	21	Red.
	22	Violet.
III.	24	Blue.
	25·5	Peacock.
	27	Green.
	29·5	Yellowish Green.
	31	Rose.
	32·5	Crimson.
	33	Purple.
IV.	34·5	Violet.
	36	Peacock.
	38	Green.
	40	Yellowish Green.
	44	Rose.
V.	48	Pale Green.
	52	Pale Rose.
	55	Rose.
VI.	60	Pale Peacock.
	64	Pale Rose.
	66	Rose.
VII.	71	Pale Green.
	74	Pale Rose.

Now the reason for this peculiar succession of tints arises from the overlapping of the successive orders for the waves of different colours. When produced as Newton made them, by the interference of light by re-flexion from the upper and lower surfaces of a film of air, there would be found, at a particular distance from the centre a particular thickness of film such that the light reflected at the second surface is exactly half a wave out of step with the light reflected at the first surface. At this place—the air film being here of a thickness equal to a quarter of the wave-length of that kind of light—that particular kind of light would be cut off by self-interference. For example, yellow light having waves 22 millionths of an inch long, would be cut off by interference when the film is $5\frac{1}{2}$ millionths of an inch thick. But as all the waves have, to begin with, lost half a wave-length (as evidenced by the central spot being black), by reason of the second reflexion being an external one, the result is that all round the centre, at such a distance that the film is $5\frac{1}{2}$ millionths of an inch thick, yellow light is reinforced, and there is seen a bright ring—the first order for yellow light. As red waves are 27 millionths of an inch long the ring for the first order for red light will be at a place where the film is about $6\frac{3}{4}$ millionths of an inch thick. Newton's rings then will seem of different sizes in different kinds of light; and since white light consists of all different colours mixed up, the Newton tints will be produced by the overlappings of all the different tints. This may be made plainer by considering a diagram (Fig. 96) in which the various sizes of waves are represented to scale. Let the distances measured horizontally from the left side represent the distances from the centre of the system of Newton's rings, the air-gap being supposed to widen in

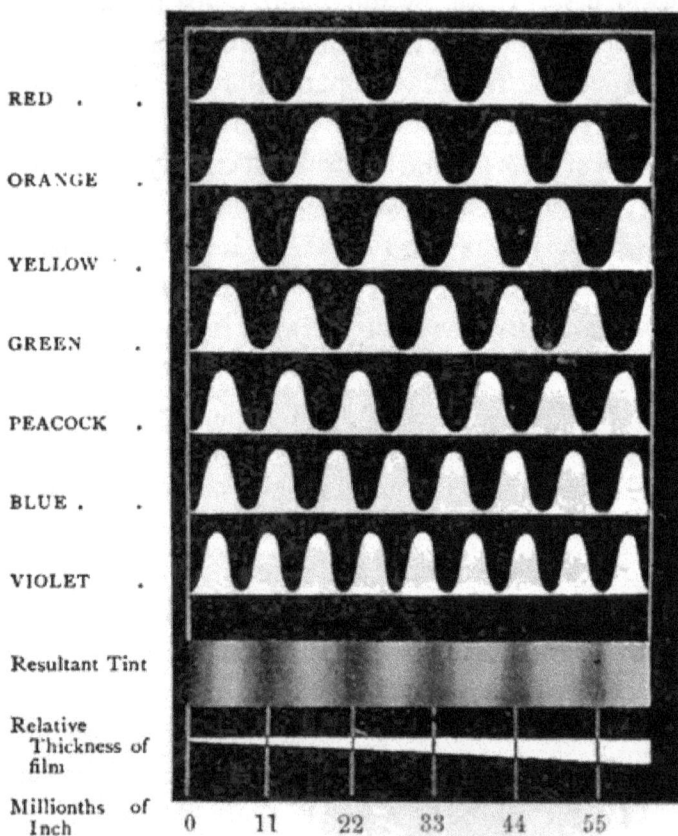

RED

ORANGE

YELLOW

GREEN

PEACOCK

BLUE

VIOLET

Resultant Tint

Relative Thickness of film

Millionths of Inch   0   11   22   33   44   55

FIG. 96.

proportion to the radius. Then if red light was the only kind falling on the apparatus it would show a system of red rings with a black centre, the distance of each successive red ring from the centre corresponding to the places where the crests of the red waves occur in the highest row. Similarly, if yellow light alone were present, there would be a rather smaller system of yellow rings seen, spaced out as are the crests of the yellow waves in the third line. And so forth for other colours. Now since the rings in the self-produced light of any one colour are of a different size from the rings in the set produced by light of another colour, it follows that when white light is used, the sets of rings of the different colours of which white light is compounded will overlap one another. At any given distance from the centre the resultant light will be the sum of all the various amounts of coloured lights at that distance. Take the rows of waves in Fig. 96 and treat them as if Fig. 96 were an addition sum of which we had to write down the total from left to right at the bottom. At the very beginning, on the left, there is nothing to add up, because the waves have not yet more than begun to rise. A little farther along all the waves are rising. Consider a distance such that the yellow wave is at its highest point. Imagine a vertical line drawn through the top of the first yellow wave. How much of the other kinds of light are present? There is a great deal of orange and some red, a great deal of green and pea-cock, some blue and some violet. Now all these added together will make a nearly white light, but rather yellowish owing to the preponderance of yellow. The result is that at the corresponding distance from the centre there will be a yellowish white ring: and the air film at this place is about $5\frac{1}{2}$ millionths of an inch thick. Now consider a distance twice as great from the centre of the rings, or at the end of the first yellow wave, where the film is about 11 millionths of an inch thick. There is no yellow, but there will be a very little orange and some red; there will be next to no green or peacock, but there will be a little blue and much violet. These colours add up to a dark purple tint, which in the coloured diagram on the screen is set down as the total in the bottom line. It may be new to you to think of adding up colours and putting down the totals, but that is the way to reckon out the resultant tint. Referring back to the table of Newton's tints we now see that they range themselves in regular orders with the purple transition tint at the end of the first order where the air film is 11 millionths of an inch thick, another purple tint at the end of the second order where the film is 22 millionths thick, and a third at the end of the third order, where the film is 33 millionths thick. In fact these darkest tints in the series correspond to the thickness at which interference occurs for yellow light.

Now these Newton's tints—produced by interference and overlapping—are in general the same as the tints which result from the introduction of our thin slices of crystals into the polariscope. And the reason why the thin slices of crystal when examined by polarised light give the same general series of tints is as follows.

Suppose the polarised light to fall on a thin slice of crystal that has its axis set obliquely across the beam. Then the vibrations in going through the crystal are split, as I have explained, into two parts, one vibrating parallel to the axis, the other part at right angles: and they do not take the same time to traverse the crystal film because of the difference between the rigidity in the

two directions. And as the wave-lengths of different
colours are different, the waves of various colours, though
they traverse the same actual thickness, emerge in
different states. When the two components of a wave
of any given colour recombine on emergence, they will
recombine to form a vibration in some new direction,
and that resultant direction—whether oblique or elliptical—will be different for different colours. Hence it
follows that the analyser will cut off more of one colour
than of another; and the light which comes through
the analyser will be the total of all the resolved parts
of each kind of light. If, for instance, the thickness of
the crystal is such that the yellow light on emerging
recombines to form a nearly vertical vibration, then the
analyser when horizontal will cut off the yellow, and the
resultant light that comes through—the total of all the
other parts—will be of a dark violet hue. Just as the
colours in the Newton's rings depend on the thickness
of air film between the glasses, so do these colours of
the film of crystal in the polariscope depend on the
thickness of the film (see p. 139).

The next object to be shown you is a slice of selenite
that is exceedingly thin at one end, and thick at the
other, being tapered as a very thin wedge. It exhibits
most magnificently the Newton's tints up to the end of
the third order. It will serve as a standard of com-
parison for other slices. For instance, there is now
placed in the apparatus a uniform piece of crystal. It
shows in the dark field the red of the second order. I
therefore know that it is precisely of the same thickness
as the wedge is at the place where the wedge shows

the same red tint. This red turns to a vivid green
when the analyser is rotated so as to give the bright
field. So again a slice which in the dark field gives a
violet of the second order changes in the bright field to
the complementary primrose tint.

I now take two prepared pieces of mica, which will
be exhibited to you first separately and then together.
One of them shows the blue of the second order, a tint
which by reference to the table is the same as that produced by an air film 13 millionths of an inch thick.
The other shows a yellow of the second order, corresponding to an air film 18 millionths of an inch thick.
Now guess what will happen if they are both put in together. Will blue and yellow make green? Not by
any means. If superposed (with their axes both at 45°
to the right) they will have the same effect as a piece of
mica would have if its thickness was equal to that of
the two added together : or it will act as a film of air in
the Newton's rings 31 millionths of an inch thick,
giving a tint which, by the table, you see to be a rose

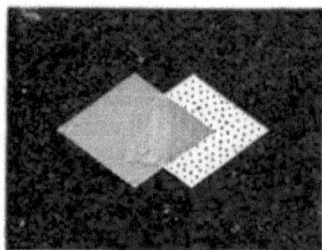

red. My assistant slides one
crystal over the other (Fig.
97) and you observe in this
case the unexpected, though
predicted, result that blue and
yellow added in this way make
pink. Let one of the crystals
now be turned about so as to

put its axis 45° to the left, so that it will act negatively,
giving the same result as if we had subtracted one thickness from the other. What tint ought it to give? Sub-

tracting 13 millionths (blue) from 18 millionths (yellow),
we obtain the answer that it ought to give the same tint
as an air film 5 millionths of an inch thick, which tint is
a grayish white. Look for yourselves and see how on
the screen where the blue (reversed) crystal overlaps the
yellow crystal, the resultant tint is a grayish white.

The next object is a wedge combination made of
twenty-four very thin pieces of mica set to overlap one
another, so as to form a wedge in steps. It, like the
smooth wedge of selenite, gives the Newton tints of the
first three orders; in this case, however, not gradating
finely into one another, but presenting sudden changes
from the tint of one thickness to the tint of the next.
Where the crystal shows the nearest approach to white,
namely, at a point half way along the first order, it corresponds to an air film having a thickness of one-half of
the length of the wave of yellow light. Hence such a
crystal is called a half-wave plate. If it is placed (with
axis also at 45°) upon one of the other slices of crystal
in the polariscope, it is observed to raise all the tints by
an amount corresponding to an addition of 11 millionths
of an inch to the corresponding thickness of air film,
and changes each to almost exactly [1] its complementary
tint. Whitish in the dark field, it is nearly black—a
very dark purple—in the light field. The next slide to
be put in the polariscope is an illustration of this
principle. In the dark field you see a white swan which,

---

[1] Not precisely exactly, as it cannot be an exact half-wave plate
for all different colours. It is selected so as to be an exact half-wave
plate for that tint that is brightest to the eye, namely, yellow, or
yellow-green.

when I turn the analyser to give us the bright field, changes to a black swan. The space, in the slide, within the outline of the swan, is covered with a piece of half-wave selenite. I possess another slide in which a baker's boy attired in white clothes, with a sack of flour on his back, changes to a chimney-sweep bearing a bag of soot.

A slice of crystal half as thick as the half-wave plate is called a quarter-wave plate. It produces a retardation of one quarter of a wave [1] (for yellow light) between the two components of vibration that traverse the slice.

Here is a beautiful object. A thin slice of selenite has been ground so as to be hollow on one face like a concave lens, thin in the middle, thicker at the edges. As a result it shows Newton's rings in a far more splendid manner than Newton's delicate air film ever showed them. Along with this concave selenite I now introduce a slide made up of twelve sectors of quarter-wave crystal, set with their axes alternately at 45° to the right and 45° to the left. They seem to dislocate the Newton's rings, pushing the alternate segments in or out by one quarter of a whole "order" of tints. Beyond these objects, and next the analyser, I now introduce upright [2]

---

[1] The quarter-wave, if set with axis at 45°, to produce as mentioned a difference of phase of a quarter between the two components, produces circularly polarised light. In all positions of the analyser light still comes through, nearly equally; there is no dark field. Also, if a quarter-wave is placed, along with other polarising objects (such for example as the concave selenite next described), but is set with its axis vertical, instead of at 45°, and is inserted between the polarising object and the analyser, it produces great varieties of tint, each tint in changing to its complementary going, while the analyser is turned, through an intermediate series of tints.

[2] See preceding note.

a quarter-wave plate.   And now, on rotating the analyser
we have the strange appearance of all these dislocated
rings of colour marching inwards to disappear at the
centre, though succeeded in turn by other rings.   Revolving the analyser in the opposite sense causes the
rings to seem to grow at the centre and march outwards.

Here is another object of great beauty, a butterfly in
form, cut out of selenite ; and here also is a heart's-ease ;
and here some daisies which, though pale yellow in the
dark field, turn to purple Michaelmas daisies in the
bright field.   To pretty devices like these there is no
end.

We may now apply our knowledge to a further study
of complementary and supplementary tints.   A few
minutes ago I showed you (Fig. 87, p. 125) how the
double-image prism as analyser gives us two polarised
images, which, when the polarised light passes through
a circular aperture of suitable size, overlap one another.
If with the same arrangement I cover the aperture with
a thin slice of mica or selenite, and superpose a vertical
quarter-wave plate, our two overlapping disks on the screen
are seen to give us two complementary colours, as though
we had two analysers, one of which had been turned
through 90°.   As we turn the double-image analyser the
two images revolve around one another exactly as they
did previously.   But as they revolve they change their
colour in regular successions.   And in every position,
whatever tint one image shows, the other shows the complementary tint ; while in every position the patch of
light where they overlap is white.   But that is because
we take the white light of the electric lamp and are

resolving it into two complementary constituents. Now if I interpose a piece of coloured glass to colour the beam, we shall resolve that coloured light into two constituents which we may describe as "supplementary" to one another. Blue glass, as we found last lecture, lets some green and violet pass as well as blue; and here again you see the fact revealed. Red glass on the contrary is fairly monochromatic; for though we split its light into two supplementary beams, both are red, scarcely differing in hue from one another.

Natural crystal patterns, produced on glass by pouring upon it some crystallisable solution which is then allowed to dry, form objects of great beauty. Here, for example, is a plate prepared from a solution of anti-pyrin. It produces an effect like frost on the window-pane. But the delicate traceries and plumes, when placed in the polariscope, show the most gorgeous play of colours, as you see. And here are some crystals of sulphate of copper, and some of pyrogallic acid, which are equally curious.

Now there are in nature sundry substances besides crystals which possess different rigidities in different directions. Thin slices of wood, for example, and bone, and horn, and many other animal and vegetable structures. Here is a thin slice of horn. It is nearly transparent and colourless. But if put into the polariscope, with its grain inclined at 45° to the vertical, you see at once the remarkable streak of colour which it produces. Here again is a quill-pen flattened out and mounted as a polariscope object. It is really quite gorgeous in its hues.

Here is a very interesting object, the natural lens from the eye of a codfish. Having a fibrous and radial structure it shows a black cross in the dark field.

Glass, under ordinary circumstances, is devoid of any difference in rigidity between one direction and another. Nevertheless, if it is suddenly heated or suddenly cooled, the unequal expansion of its parts produces differences in rigidity which make themselves visible in the polariscope. Here, for example, is a small square piece of glass, which at present shows no colour or any other effect when placed in the polariscope. But if it is dropped into a heated brass frame which will quickly warm its edges before the central part of it has time to expand, its structure will be put under unequal stresses, and the resulting strains will show themselves in the strange patterns of colours which you now see growing into sight upon the screen.

If a hot piece of glass is suddenly chilled, so that the outer part cools and contracts before the inner part has time to cool, the piece may acquire and remain in a state of permanent strain. Such glass, usually described as "unannealed," is very liable to break[1] with almost explosive violence. Here is a square piece of glass, of no colour in itself, but which has been suddenly chilled. Its state of permanent strain is at once revealed by the peculiar pattern and the dull tints that seem to form around a distorted cross radiating from its centre. A

[1] The extreme case is presented by "Rupert's drops," which are drops of melted glass suddenly cooled by dropping them into water. When examined in polarised light (best when immersed in a small glass tank filled with oil to obviate surface reflexions) they show fine colours.

second square of glass which has been still more suddenly
cooled shows the same black cross (Fig. 98), but the tints

FIG. 98.

in the corners of the square run
up into the second order. Here
again is a short cylinder of glass
which was suddenly chilled all
round its outside. The peripheral
surface has contracted upon the
inner part and compressed it with
an enormous force. As a result
you see not only the black cross
indicative of a radial disposition of the axes of elasticity,
but a number of concentric rings coloured with the now
familiar succession of Newton's tints right up to the
fourth order.

And now I am going to squeeze a piece of glass
mechanically, by gripping it in a strong brass frame and
then forcing a point against its side by turning a strong
screw. In the dark field the glass itself shows neither
light nor colour, until I put on the screw. But so soon
as compression is applied a luminous pattern at once
seems to grow, stretching off in two patches at about 45°
on each side of the point where the screw point has been
forced against the glass. Tightening the screw makes
the internal strain greater, and the pattern more brilliant.
Loosening the screw releases the strain, and the glass
resumes its ordinary colourless state. So, you see, you
can use polarised light not only to detect false gems from
real, not only to tell glass from crystal, but also to
ascertain whether any piece of glass is likely to break or
not. Any piece of glass that has been too suddenly

cooled, that is, has not been properly annealed by slow
cooling down from the furnace heat, can always be
detected by the colours it shows when placed in the
polariscope between polariser and analyser.

For such purposes a very simple polariscope, such as
any ingenious boy might construct for himself, is quite
sufficient.   Here (Fig. 99) is such a polariscope, made
entirely of glass.   The polariser is simply a flat piece of
window-glass, 9 inches long by 5 inches wide, blackened

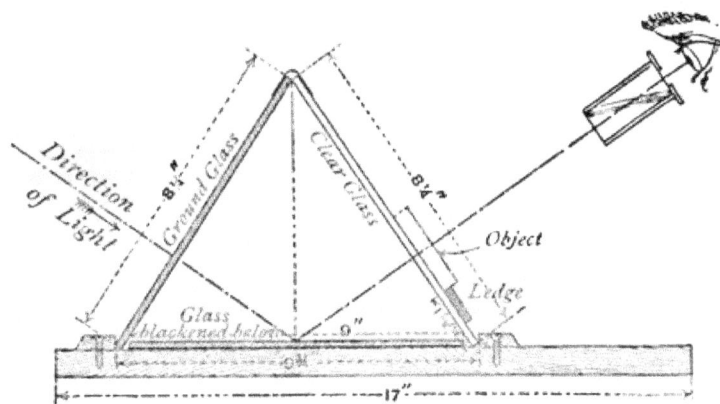

FIG. 99.

with black varnish on its under side, and laid down on a
simple frame of wood.   Two other pieces of glass are
cut of the size 8¼ by 5 inches.   One of these is of clear
window-glass, the other of ground glass.   Across the
lower part of the clear piece, and at $1\frac{5}{8}$ inch from its
edge, is cemented a strip of glass 5 inches long by
1 inch broad, to serve as a ledge on which to support
the objects when being looked at.   These two pieces of
glass are joined at the top by a hinge of paper or cloth
cemented to them ; and they stand up, like a roof, over

the piece of blackened glass, being kept in their places on the baseboard by two strips of wood which are fastened to the board $9\frac{3}{4}$ inches from one another. The baseboard should be 17 inches long by 5 inches wide. Daylight or lamp-light is allowed to strike upon the ground glass, and thence passes down to the blackened glass, is reflected at an incidence of about 57° to its surface, and so passes as a polarised beam through the piece of clear glass on its way to the eye. As analyser, seeing that Nicol prisms are expensive, a cheap substitute must be found. One that is quite good enough for many purposes, may be made by taking a bundle made of eight or ten very thin slips of glass,[1] each about $1\frac{1}{2}$ inch long and $\frac{5}{8}$ inch wide, and fixing them with sealing wax obliquely across a small wooden tube or box with open ends. They should be fixed in the wooden tube so that the glass slips are inclined at about 33° to the direction in which the light is to pass through the tube.

With quite simple apparatus you can verify and repeat many of the experiments that have now been shown before you. There are many others, equally beautiful, that I have not shown ; for in a single lecture one can only deal very incompletely with this fascinating and complicated subject of polarisation. I have not shown you how quartz crystals possess a special property of rotating the polarised light, nor have I told you how solutions of sugar and sundry other liquids are found also to produce

---

[1] The very thin glass used for "covers" for microscopic objects is suitable. It is usually supplied only in round cover-disks. But any good optician could procure rectangular slips of the size named.

an optical rotation. Indeed, the regular way adopted in sugar factories to measure the amount of sugar in a watery syrup is to put some of it into a polariscope and measure how much it turns the direction of the vibrations. Lastly, I have said nothing about the remarkable discovery with which Faraday crowned his researches in this place, namely, that the polarised waves of light can be rotated by a magnet. Let me hope that some day you may learn of these marvellous discoveries, to which the things you have seen to-day constitute a first step.

APPENDIX TO LECTURE III

THE ELASTIC-SOLID THEORY OF LIGHT

ON p. 34 it is remarked that light-waves travel slower in denser media; and on p. 129 it is explained how in a double-refracting crystal the waves are split into two sets which travel with different velocities. It is expedient to enter further into the question of the velocity of propagation of light-waves. If it is assumed as a fundamental point that the velocity of propagation of a wave is equal to the square-root of the elasticity of the medium divided by its density (or, as expressed in symbols, $v = \sqrt{E \div D}$, which is Newton's law), then it is only possible to account for the co-existence of two different velocities by supposing that displacements in different directions either evoke a different elasticity or call into operation a different density. But, since the medium of which the waves constitute light is the ether, one has to deal, in the case of the transmission of light through crystals, with the ether as it exists in the crystal. If we assume that the ether acts as an incompressible homogeneous elastic solid, then the ordinary theory of elasticity suffices as a theory of the ether. For long this " elastic-solid theory " of the ether has held sway, and has received elaborate mathematical treatment at the hands of Green, Fresnel, MacCullagh, Neumann, Cauchy, and others. On this view the ether particles within crystals are arranged differently in different directions, symmetrically with respect to three rectangular axes, and therefore the properties of the ether as a medium for transmitting waves will be modified by the presence of the crystalline matter.

But here a difference of view may arise; for it may be held (with Fresnel) that the presence of the crystalline matter modifies the density of the ether without altering its elasticity; or it may be supposed (with MacCullagh and Neumann) that the presence of the crystalline matter modifies the elasticity in different directions without affecting its density. In either case the assumptions lead to equations that fit the fundamental facts of double-refraction and polarisation. But there arises this difference that whereas the theory of Fresnel supposes the displacements to occur at right angles to the so-called "plane of polarisation," that of MacCullagh treats them as executed in that plane. As to the actual direction in which the displacements are executed, the properties of tourmaline suffice (apart from other proofs) to determine the fact that in the extraordinary wave, which is transmitted, the displacements are executed parallel to the principal axis of the crystal, while in the ordinary wave, which is absorbed, the displacements are at right angles to that axis. The simple proof being (see *Philosophical Magazine*, August 1881) that tourmaline is opaque (at least in thick slices) to all light travelling along the principal axis of crystallisation; hence it absorbs those vibrations which are transverse to that axis. (Compare p. 119 above.)

But the elastic-solid theory is not the only possible theory of light. Instead of supposing the ether to be itself modified in arrangement or properties by the presence of crystalline matter we might suppose it to be itself isotropic, having equal elasticity and density in every direction, but that in its motions it communicates some of its energy to the particles of matter through which the wave is travelling. If the particles of gross matter thus load the ether their vibration will during the passage of the wave take up some of the energy and retard the rate at which the group of waves can travel. We should then have a difference between the velocity of propagation of the individual waves themselves and the velocity of propagation of the group of waves; and in that case the velocity of propagation of the group—the apparent velocity of light—would be slower

than the velocity of the waves themselves, and would be
different for waves of different frequency. This is in fact
the phenomenon of *dispersion.* In the case of crystalline
media the retardation and the dispersion would be different
in different directions, and would depend upon the direction
of the displacements with respect to the axes of the crystal.
But as to the connection between the molecules of matter
and the ethereal medium involved in such theories, very ·
little is known, and there is room for many different
hypotheses as to the nature of such connection. Helmholtz,
Kelvin, Lommel, Sellmeier and others have made various
suggestions of which an admirable account is to be found
in Glazebrook's "Report on Optical Theories," *British
Association Report,* 1885.

The electromagnetic theory of light which Maxwell
founded upon the basis of the experimental work of Faraday
has now definitely superseded all the purely mechanical
theories. Some account of this theory is given in the
Appendix to Lecture V. (p. 230).

The only other poi that need claim attention here is the
use of the term "plane of polarisation." This term, which
is variously defined by different writers, is used to denote a
plane with respect to which the polar properties of the
wave can be described. It must necessarily contain the
line along which the wave is being propagated (*i.e.* the
"ray" lies in this plane); but, so far as the orientation of
this plane around the ray is concerned, its definition with
respect to the polar properties is purely a matter of convention. The following is Herschel's definition (*Encyclopedia
Metropolitana,* article "Light," p. 506)—"The plane of
polarisation of a polarised ray is the plane in which it must
have undergone reflexion, to have acquired its character of
polarisation; or that plane passing through the course of
the ray perpendicular to which it cannot be reflected at the
polarising angle from a transparent medium; or, again,
that plane in which, if the axis of a tourmaline plate exposed
perpendicularly to the ray be situated, no portion of the ray
will be transmitted." If we refer to the Nicol prism (Fig.
84, p. 123) we shall see that, according to the convention

thus laid down by definition, the plane of polarisation of the
light that emerges is parallel to the longer diagonal of the
end-face ; and the vibrations are executed at right angles to
this.   To avoid periphrasis in these Lectures the author
speaks of the plane in which the vibrations are executed as
the plane *in which* the wave is polarised (see descriptions
of Figs. 69-73, pp. 113-117).

LECTURE IV

THE INVISIBLE SPECTRUM (ULTRA-VIOLET PART)

The spectrum stretches invisibly in both directions beyond the visible part—Below the red end are the invisible longer waves that will warm bodies instead of illuminating them—These are called the calorific or *infra-red* waves. Beyond the violet end of the visible spectrum are the invisible shorter waves that produce chemical effects—These are called actinic or *ultra-violet* waves—How to sift out the invisible ultra-violet light from the visible light—How to make the invisible ultra-violet light visible—Use of fluorescent screens—Reflexion, refraction, and polarisation of the invisible ultra-violet light—Luminescence : the temporary kind called Fluorescence, and the persistent kind called Phosphorescence—How to make " luminous paint" —Experiments with phosphorescent bodies—Other properties of invisible ultra-violet light—Its power to diselectrify electrified bodies—Photographic action of visible and of invisible light—The photography of colours—Lippmann's discovery of true colour-photography—The reproduction of the colours of natural objects by trichroic photography—Ives's photochromoscope.

ALL kinds of light in the visible spectrum are comprised between the extreme red at one end and the extreme violet at the other. Their wave-lengths vary between about 32 millionths of an inch (extreme red) and 15 millionths of an inch (extreme violet). But besides the waves of various colours, between those limits, which are visible, there are other waves that bring no sensation to our eyes, which are invisible, and yet are light-waves. In brief, the spectrum extends in both directions invisibly, both below the extreme red and beyond the extreme violet.

Perhaps you raise the objection that if such waves are invisible they cannot be waves of light. Well, if you were to lay down as a definition beforehand that the term "light" must be applied only to the waves that are visible to the human eye, there is nothing more to be said. But what if there are other eyes, or other processes that will enable these waves to be observed? Further, if it is found that these invisible waves agree with the visible waves in other important respects, if, in fact, it is found that they can be reflected, refracted, polarised, and diffracted, then we are bound to regard them as *light*. They may have wave-lengths that are larger than that of the red waves, or smaller than that of the violet waves, and so our eyes, with their limited range of perception, may fail to be sensitive to them. Nevertheless if in their physical properties they agree with the visible kinds, then the fact that to us they are invisible simply demonstrates the imperfection of our eyes. Had we lived all our lives behind screens of red glass we should never have known anything of green or blue waves : we should have been blind to waves of these particular kinds. But though we should never have seen them that would not prove that they were not waves of light.

Now that part of the invisible spectrum which consists of waves of too large a size—of too great a wave-length—to affect our eyes possesses another property, namely, that of warming the things upon which it falls. Some of the visible waves, particularly those toward the red end of the spectrum, share the same property, but to a less extent. The longer invisible waves are called variously the *calorific* or *infra-red* waves. We shall deal with these in the next lecture. At the other end of the spectrum, beyond the violet, we have again waves which are invisible by reason of being of too small a size to affect our sense of sight; and these possess several remarkable properties. They are active in producing certain chemical effects, notably those known as photo-chemical or photographic. They produce certain physiological effects also on animal and vegetable tissues. They actively provoke in certain bodies the property of shining in the dark, or phosphorescence. Lastly, they have certain electrical properties. These short waves are known by the various names of *actinic*, *photographic*, or *ultra-violet* waves. The last of these terms is much to be preferred. Some of these chemical effects are also produced by visible light, especially by the blue and violet waves. Fig. 100 is a diagram which gives a general idea [1] of the distribution of these effects for waves of different lengths. The greatest luminosity to the eye is possessed by waves having a wave-length of about 22 millionths of an inch or 0·00055 of a millimetre. The greatest heating effect occurs with waves of about 40 millionths of an inch, or 0·001 of

---

[1] A table of wave-lengths and frequencies of all kinds of light from the lowest infra-red up to the highest ultra-violet has been added as an Appendix to this Lecture.

a millimetre. The greatest chemical [1] effect occurs with waves of about 16½ millionths of an inch, or about 0·00041 of a millimetre.

Now, it is desirable for purposes of experiment to separate the waves which can produce one of these effects from those which produce another. If we desire to sift out the ultra-violet waves from all other kinds, there are several courses open to us. Firstly, we may

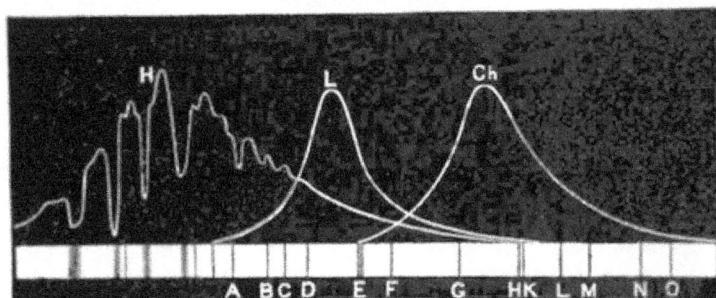

FIG. 100

use prisms which will disperse the waves and sort them out into a spectrum according to their sizes. Secondly, we may accomplish the same thing by using a diffraction grating to produce a spectrum. Or, thirdly, we may employ, as a means of sifting, sheets of different substances that have the power of absorbing waves of one sort while transmitting those of another. This last process we found excellent when applied to visible light,

[1] This is on the assumption that the effect is measured by a particular chemical reaction, viz. the darkening of chloride of silver. If a different reaction, say, for example, the darkening of ferro-prussiate salts ("blue-prints") were taken as a basis of measurement, the maximum effect would be found to occur at some other point in the spectrum.

for by using a red-coloured glass we were able to cut off
all the other colours and leave only the red.   Unfortunately no perfect filter-screen exists that will cut off all
the visible light and yet transmit the ultra-violet waves.
Glass tinted a deep violet colour with manganese, or with
manganese and cobalt, may serve to cut off most of the
visible light while transmitting a fair proportion of ultra-
violet waves, mixed with some violet light.   For many
purposes this is good enough.   But, unfortunately, every
kind of glass cuts off the extreme part of the ultra-violet
light.   Even the lightest crown glass, though moderately
transparent to waves from 15 millionths to 11 millionths
of an inch long is totally opaque to all waves smaller than
11 millionths ; while dense flint glass (containing lead)
is opaque to everything beyond the wave-length of 13·3
millionths of an inch.   Hence, for experiments on ultra-
violet light it is expedient not to use glass lenses or
prisms, provided some more transparent medium can be
found.   Happily both quartz and fluor-spar are much
more transparent to ultra-violet waves than glass is.
Quartz transmits them down to about 8·1 millionths
of an inch, and fluor-spar down to 8 millionths.   My
lantern is on this occasion provided with condensing
lenses of quartz.   When we want a spectrum we will
use a quartz prism and a focusing-lens also cut from
quartz crystal.

Let me now proceed to demonstrate some of the
photographic properties of light-waves.   Here is a piece
of ordinary " printing-out " paper, that is paper which
has been covered with a sensitive film impregnated with
chloride or bromide of silver, which, when exposed for

a sufficient time to light, turns nearly black. Over this sheet of sensitised paper I place some stencil-plates cut out in sheet-zinc ; and then expose it to the white light that comes from an electric arc-lamp on the table. In half a minute the paper will have darkened sufficiently for you to see that where the light-waves have fallen upon the exposed parts they have produced the chemical action, and have printed the patterns of the stencils. In this experiment all kinds of rays—calorific, visible, and actinic—have been allowed to fall on the paper ; but which of them were the agents in producing the effect? That is easily tested. We turn on the light in the optical lantern, using the quartz lenses and prism to produce a spectrum for us. Then along the whole length of the visible spectrum, and to a distance into the invisible spectrum at both ends, we stretch out a long strip of sensitised photographic paper. It must be left there for several minutes, during which time we may investigate another point.

Here is another sheet of sensitised paper. Over it I lay a sheet of opaque tin-foil, through which there have been cut a number of holes. Over these holes are laid a number of thin slices of various materials : (1) window glass ; (2) flint glass ; (3) red glass ; (4) green glass ; (5) blue glass ; (6) quartz ; (7) fluor-spar ; (8) rock-salt ; (9) ebonite. I now slide the whole arrangement under the beams of the arc-lamp, which is set to throw its whole light downwards. If any of these materials cuts off the active waves we shall find it out at once, for the paper will be darkened only under those sub-stances that are transparent to the photographic rays.

Two minutes' exposure suffices for our simple test. On bringing out the sheet you will note that ebonite (which is black) and red glass have alike stopped off the whole of the photographic rays. Green glass has stopped off the greater part of them, and the flint glass has evidently not transmitted them all. But under the blue glass, the quartz, the rock-salt, the fluor-spar, and the window glass the paper seems to have darkened about equally. With a more refined test we should discover differences between these also. One fact we have proved, which is of practical importance, namely, that red light does not affect a photographic film though it affects our eyes. Every photographer knows this ; for he takes advantage of it in using ruby glass or red-coloured tissue to cover the windows of his "dark-room," or to screen the lamp by whose light he works in preparing and developing his plates.

Meantime our long strip of sensitised paper has been exposed to the spectrum, and now, examining it, we see that it has sensibly darkened at the violet end and beyond the end of the visible violet to some distance into the region where our eyes see nothing; in short, the photographic spectrum lies mostly beyond the violet, the most active waves being shorter than any that are visible. But we must not forget that with other chemical preparations the range of sensitiveness can be changed. To Captain Abney science owes the introduction of emulsions of bromide of silver in films of gelatine, prepared in such a way as to be sensitive not only to violet light or ultra-violet, but also to green, to yellow, and even to red waves.

Another chemical effect which light-waves can produce is to cause mixed hydrogen and chlorine gases to
enter into combination. These gases (prepared by
electrolysis of hydrochloric acid) may be kept mixed,
but not chemically combined with one another, in glass
bulbs for any length of time, provided they are kept in
the dark. If exposed to the diffused light of a room
they slowly combine. But if exposed to direct sunlight
or to the light of the arc-lamp they combine with extraordinary violence and explode the bulb. Again, the
question arises: which part of the light is it that produces
the effect? Certainly not the red waves, for these bulbs
of mixed gas may be exposed freely if protected by red
glass and will not explode. The active kind of waves is
in this case also the ultra-violet kind.

A thin glass bulb containing the mixed gases is now
taken by my assistant
from a tin box, where
it has been kept in the
dark. To prevent acci-
dents he places it in an
empty lantern (Fig. 101),
into the nozzle of which
we will direct, from out-
side, the beams of an
electric arc-lamp. To
cut off the bulk of the
ordinary light I inter-

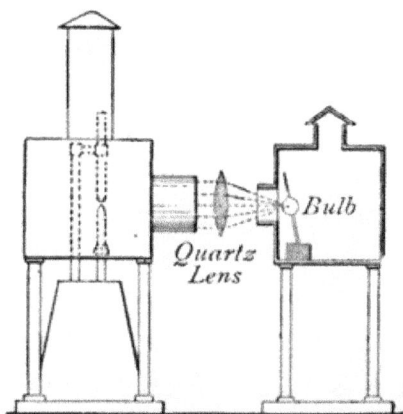

FIG. 101.

pose first a sheet of violet glass, which allows only violet
and ultra-violet to pass. Then, interposing a quartz
lens, I concentrate the beam upon the bulb, when—bang

—it explodes, demonstrating the activity of waves of this sort.

Perhaps it may have struck some of you that if so great a photographic activity is possessed by waves that are invisible to our eyes, it ought to be within the limits of possibility to photograph things that are invisible. And so it is. It is now some twenty years since a lecture was delivered in this theatre on the photography of the invisible by the veteran chemist, Dr. J. Hall Gladstone, who succeeded in photographing images of things quite invisible to the eye. Behind me, against the wall, stands a drawing-board covered with a white sheet of cartridge paper. The light of the arc-lamp shines on it. You see merely a white surface. The photographer, Mr. Norris, has brought his camera here and he will now take a photograph of it. When he develops the photograph you will find that the photograph will reveal the fact that an inscription has been written upon the sheet, which, though invisible to you, can be photographed by the camera.[1]

Since photographic action serves to detect these ultra-violet waves, even in the absence of all kinds of visible light, it may be used in the further exploration of the properties of these invisible waves. We might apply photographic plates to prove the possibility of the re-

---

[1] The inscription was written on the sheet with colourless sulphate of quinine dissolved in a solution of citric acid. This substance fluoresces, and in the act of fluorescing destroys the ultra-violet light, which would otherwise be reflected from the parts of the paper so treated. The parts where the sulphate of quinine has been applied consequently come out in the photograph darker than the untouched surface of the paper.

flexion and refraction of these waves, as well as of their interference and polarisation. There exists, however, another and more ready method of investigation, to which we will now proceed.

Instead of photographing the invisible we may make it visible to the eye by applying the discoveries of Herschel, Brewster, and Stokes. There are a number of solid substances, such as fluor-spar, uranium glass, and also of liquids, such as petroleum, solutions of quinine, and of many of the dye-stuffs derived from coal-tar, which present the appearance of a surface-colour different from that of the interior. Thus quinine is colourless, but shows a fine blue tint on the surface exposed to the light. Uranium glass is itself yellow, but has a splendid green surface-tint. The fact is that these substances have the property of absorbing the very short waves of ultra-violet light and transforming them into waves of longer length that are visible to our eyes. To this phenomenon Stokes gave the name of *fluorescence*. Let us see a few of the principal cases.

From the optical lantern, provided for the present experiments with quartz lenses, a beam of light streams forth. Over the nozzle of the lamp is now placed a cap of dark violet glass to cut off all the visible light except a little violet that unavoidably accompanies the invisible ultra-violet waves. This beam is directed upon a cube of uranium glass; which transmutes the invisible waves into a brilliant green. And you see the glass cube standing out vividly in the darkness. I hold in the beam a bottle of paraffin oil—it seems brilliantly blue. A green decoction of spinach leaves (boiled first in water, then dried, and lastly extracted with ether) exhibits a strange blood-red fluorescence on its surface.

Here is a row of specimen bottles containing fluorescent liquids. Yellow fluorescein gives a splendid green fluorescence; pale pink eosin (made by diluting red ink) gives an orange fluorescence. A crimson solution of magdala-red gives a scarlet fluorescence; and colourless quinine gives its characteristic surface-blue.

These things may be even more strikingly shown by reflecting the ultra-violet beam down into a tall glass cylinder filled with fluorescent liquid. A quartz lens placed just above the jar (Fig. 102) serves to concentrate the beam into a sharp cone of colour. I take a second jar filled simply with dilute ammonia-water, and project the beam down it. Then I sprinkle into the water a few grains

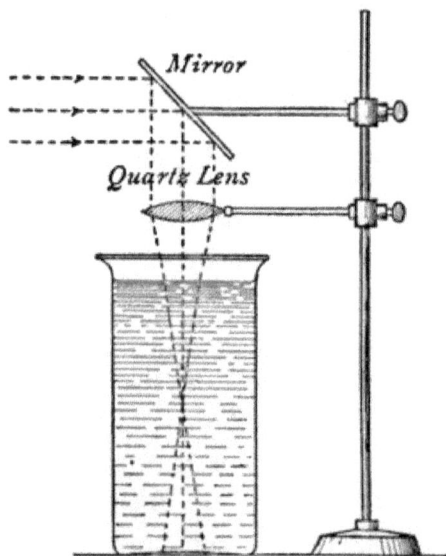

FIG. 102.

of dry fluorescein. As they dissolve there descend to the bottom curling wreaths of bright green hue and indescribable beauty. A few chips of horse-chestnut bark, or of ash bark, would yield similar effects.

And now, perhaps, you will appreciate the secret

of the photography of the invisible. The inscription painted on the white sheet was painted with a solution of quinine. You shall *see* for yourselves the invisible inscription ; for I have only to cast upon it a beam of ultra-violet light to cause the parts painted with quinine to shine out in pale blue amid the darkness.

Here are some other sheets of card on which fluorescent patterns have been painted. On one of these, side by side, are two *fleurs-de-lys*, which in daylight appear to be equally yellow. One is painted in common gamboge, the other in fluorescein. But when I turn upon them the beam of the lamp filtered through dark blue glass, the whole card looks deep violet, one of the lilies seeming black, the other luminous and greenish. Another card, when viewed in ordinary white light seems to be merely yellow all over : but as part of the yellow surface is painted in gamboge, and the other fluorescein, the effect, when examined in the ultra-violet beam is to give a black pattern on a bright ground.

Twenty years ago when the late Professor Tyndall was delivering in the United States his famous series of lectures on light, he received from President Morton, of the Stevens Institute at Hoboken, some samples of a new fluorescent hydrocarbon, "thallene," prepared from petroleum residues. Some large sheet diagrams of flowers, painted in parts with thallene and other fluorescent materials, were amongst the objects which Professor Tyndall brought back to London. These have been carefully preserved in the Royal Institution, and I am able to show you them in all their beauty.

One of them represents a wild mallow, the leaves being coloured with some substance which fluoresces green, whilst the flowers have a pale purple fluorescence. The effect of throwing on this object light that has passed through a dark blue or dark violet glass is very striking.

Of all substances, however, that are known to me, the most highly fluorescent is a rather expensive crystalline product called by chemists the platino-cyanide of barium. In ordinary light it looks like a pale yellow or greenish yellow powder, closely resembling powdered brimstone. When a piece of paper covered with this substance is held in the ultra-violet beam it emits a yellowish-green light far surpassing in brilliance that emitted by uranium glass or by fluorescein. Here is a small fluorescent screen of platino-cyanide of barium that has been in my possession some sixteen years.

Now, having so excellent a means of making ultra-violet waves visible, let us apply the fluorescent screen, as Stokes did in 1851, to explore the ultra-violet spectrum. My assistant puts up the quartz prism in front of the slit to give us once more the spectrum. Taking a long sheet of cardboard that has been painted over with quinine, I hold it so that the spectrum falls upon the middle of the prepared surface. And now you see that the spectrum stretches visibly away beyond the violet end, for here, crossed by several transverse patches of brighter light, the ultra-violet spectrum comes into view as a pale-blue extension. Substituting a sheet of uranium glass we note a similar extension visible into the ultra-violet to a distance that makes this part of the spectrum seem quite twice as long as the whole visible part. Here, best of all, is a fluorescent screen covered with platino-cyanide of barium. And now we see the "long spectrum," stretching away to three or four times the length of the visible part from red to violet. If placed at the other end, below the red, these fluorescent screens show nothing whatever. They are excited into luminous activity not by the long waves, but by the very short ones.

Let us then avail ourselves of the luminous quality of the fluorescent screen to examine afresh the different degrees in which transparent substances transmit or absorb these ultra - violet waves. The ultra - violet part of the spectrum now falls upon the screen, the surface of which is thereby stimulated into emitting its fine greenish light. Across the path of the invisible beam I interpose a piece of window glass. The light is dimmed but not extinguished. A piece of flint glass cuts it off altogether; a piece of blue glass dims it, but does not cut it off; while a

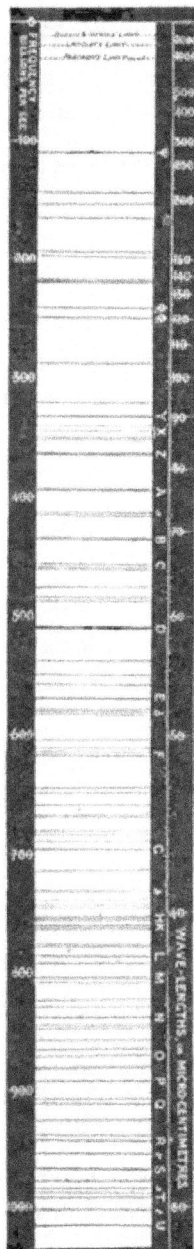

Fig. 103.

piece of red glass proves to be absolutely opaque.
A slice of quartz crystal is fully transparent; one
of calc-spar rather less so, while a thin film of yellow
gelatine is quite opaque. These experiments con-
firm those we made by the use of photographic
paper.

And now in a very few moments we can demonstrate
reflexion and refraction of the ultra-violet waves. I
place my fluorescent screen in a position where none of
the waves fall upon it. Then holding a mirror in the
invisible beam I reflect ultra-violet waves upon the
screen, which at once shines with its characteristic
greenish tint. To prove refraction I interpose in the
invisible beam a quartz prism, which deviates ultra-violet
waves upon the fluorescent screen, and again it shines.
Polarisation may be proved by using two Nicol prisms
precisely, as was done in my last lecture for ordinary
light.

This phenomenon of fluorescence is only one of a
number of kindred phenomena, now generally classified
together under the name of *Luminescence*. This name
was given by Professor E. Wiedemann to all those cases
in which a body is caused to give out light without
having been raised to the high temperature that would
correspond to the ordinary emission of light. To make
ordinary solids red-hot they must be raised to between
400° and 500° of the centigrade scale of temperature.
To make them white-hot—that is to say, to cause them
to emit not only red, orange, and yellow, but also green,
blue, and violet light, they must be raised to 800° or
1000° of temperature. At red-heat a body emits few or

no green, blue, or violet waves. But as we have seen in the examples of fluorescence some substances while quite cold can be stimulated into giving out light by letting invisible ultra-violet waves fall upon them. So we may well inquire what other cases there are of the emission of cold light. Accordingly, on the table before you there are enumerated the various cases of luminescence.

Phenomenon.	Substance in which it occurs.
1. Chemi-luminescence . .	Phosphorus oxidising in moist air ; decaying wood ; decaying fish ; glow-worm ; fire-fly ; marine organisms, etc.
2. Photo-luminescence :	
(*a*) *transient* = Fluorescence	Fluor-spar ; uranium - glass ; quinine ; scheelite ; platino-cyanides of ·io·· bases ; eosin and many coai-tar products.
(*b*) *persistent* = Phosphorescence	Bologna-stone ; Canton's phosphorus and other sulphides of alkaline earths ; some diamonds, etc.
3. Thermo-luminescence . .	Fluor-spar ; scheelite.
4. Tribo-luminescence . .	Diamonds ; sugar ; quartz ; ura yl nitrate ; pentadecyl-paratolylketone.
5. Electro-luminescence :	
(*a*) Effluvio-luminescence .	Many rarefied gases ; many of the fluorescent and phosphorescent bodies.
(*b*) Kathodo-luminescence .	·Rubies ; glass ; diamonds ; many gems and minerals.
6. Crystallo-luminescence . .	Arsenious acid.
7. Lyo-luminescence . .	Sub - chlorides of alkali-metals.
8. X-luminescence . . .	Platino-cyanides ; scheelite, etc.

The Chemi-luminescence which heads the list includes those cases in which the emission of cold light is accompanied by chemical changes. The best known instance is the shining in the dark of phosphorus when slowly oxidising in moist air. Lucifer matches, if damped and then gently rubbed, shine in the dark. The best way to show this is to take a sheet of ground glass, dip it into warm water, and then write upon its roughened surface with a stick of phosphorus, which, for the sake of safety, is held in a wet cloth. See how, on lowering the lights in the theatre, the inscription I have scribbled upon the glass shines with a pale blue glimmer. In a few minutes the film of phosphorus will have oxidised itself completely, and the emission of light will be at an end. Curiously enough, this light itself consists not only of the blue waves that you can see, but of some invisible waves also, which have photographic properties, and can, like Röntgen's rays, affect a photographic plate that is enclosed behind an opaque screen of black paper. It is now known that the emission of light by glow-worms and fire-flies, and by the innumerable species of marine creatures and deep-sea fishes that shine in the dark, belongs to the class of chemi-luminescence. So does the emission of light by the microbes that are developed in decaying fish and in rotten wood. In all these cases there is chemical decomposition at work.

Under the next heading—Photo-luminescence—are included those cases in which bodies give out cold light under the stimulation of light-waves. Of this phenomenon there are two cases. Fluorescence is one,

and in that case the emission of light is temporary,
lasting only while the stimulation lasts. The other
case is that known as Phosphorescence, a term applied
to those instances in which the emission of light persists
after the stimulation has ceased. The earliest known
example of phosphorescence is that of the celebrated
Bologna stone. A shoemaker in the city of Bologna,
Casciarolo by name, about the year 1602 discovered
a way of preparing a species of stone which, after
having been exposed to sunlight, would shine in
the dark. This was done [1] by the partial calcination
of "heavy-spar"—the sulphate of barium—found near
that city. Here is a small sample on the table. Since
that time many other substances have been found to
possess the same property. Some diamonds, as Robert
Boyle observed, have this property. And amongst
artificial substances the sulphide of calcium (Canton's
"phosphorus") and sulphide of strontium possess the
property to a very high degree. Sulphide of calcium
can be prepared by pounding up oyster-shells and
heating them to redness, mixed with a little brimstone,
in a closed crucible. The addition of small quantities
of other materials—a little bismuth, or manganese, or
copper—has a remarkable influence in aiding the
production and in changing the colour of the light
emitted. The substance sold as Balmain's luminous
paint is a preparation mainly of sulphide of calcium
with a trace of bismuth. Of all these artificial phos-

---

[1] See a singular little volume published in Rome in 1680, by
Marc' Antonio Cellio, with the title *Il Fosforo, o' vero la pietra
Bolognese, preparata per rilucere frà l'ombra.*

phori the most powerful by far is a new luminescent paint prepared by Mr. Horne.

Behind me an electric lamp is arranged to throw a beam of light down a tube. At the bottom of this tube I expose to the stimulation of its beams a few of these phosphorescent stuffs. Here is the bit of Bologna stone. On removing it from the beam it shines in the dark, but not nearly so brightly as the bit of Horne's new material, the light of which is equal to about one-tenth of a candle for each square inch of surface exposed. One can see to read print by the light of a bit of this stuff. I have heard of people using a glow-worm in the same way in order to read at night. Here is a diamond ring [1] having five fine diamonds. On exposing it for a minute to the light, and then bringing it out into the darkness, two of the diamonds are seen to shine like little glow-worms.

Here is a box containing a row of glass tubes, in each of which is a white powder. These powders are phosphorescent. But first they must be stimulated, for at present they emit no light. Let us expose them for thirty seconds to the beams of the arc-lamp. On then bringing them into the darkness of the theatre it will be seen that they glow brightly in all the colours of the rainbow.

Here, again, is a large sheet of glass which has been painted over with luminous paint. I lay my hand against it, and expose it for a minute to the beams of the arc-lamp. Extinguishing the light, you see the whole sheet splendidly luminous, save where the shadow

[1] The property of Dr. J. H. Gladstone, F.R.S.

of my hand appears as a black silhouette. In this case the luminescence is at first of a fine blue tint. In a few minutes as it fades out it becomes whiter; but it will go on all night giving out a faint light, and even then will not have yielded up its whole store of luminous energy. Even after having been kept six weeks in darkness a sheet of luminous paint will still emit waves that will fog a photographic plate. If one takes a sheet of luminous paint that has been exposed to light, and of which the phosphorescence has already died away, one finds that merely warming it will cause it to shine more brightly, though afterwards it is darker. Here is such a sheet. I place my hand against the back, and you note that where my hand has warmed it it shines more brightly. If one makes the converse experiment of chilling a sheet of luminous paint while it is phosphorescing, one finds its light dimmed, but it grows brighter while being warmed. Professor Dewar has made the curious discovery, that when cooled to a temperature of about 200° below zero in liquid air, many substances become phosphorescent that are not so at ordinary temperatures. Thus he has shown in this theatre how such things as feathers, ivory, and paper become highly phosphorescent on being cooled to these low temperatures and then illuminated. They seem when chilled to acquire the power of absorbing luminous energy and storing it for subsequent emission when warmed. The analogies between these properties and those of luminous paint are most suggestive. A sheet of luminous paint which has been exposed and cooled becomes a veritable lamp of Aladdin. One has but to warm it by the hand and it shines.

Here we touch upon the third sort of luminescence named in our list (p. 175), namely Thermo-luminescence. This term is applied to the property possessed by various minerals, particularly by the green sorts of fluor-spar, to shine in the dark on being heated. Over a large atmospheric gas-burner a square of sheet-iron has been heated to near redness. Upon this hot surface, invisible in the darkness, I scatter out of a pepper-box some fine fragments of crushed fluor-spar. As they heat up they shine like little glow-worms. They shine brightly for a few minutes, then fade, but would continue for several hours to emit a faint glow. After having been once thus heated they seem to have lost their store of luminous energy, for on a second heating they do not again luminesce.[1]

The term Tribo-luminescence, which stands next on the list, relates to the production of luminescence by friction. There is a very simple experiment that can be tried at home without any apparatus. Crush a lump of sugar in a perfectly dark room. In the act of being crushed it emits a pale luminescence. So do crystals of uranium nitrate if shaken up in a bottle, or triturated in a mortar. Let me show it to you on a larger scale.

---

[1] Many other minerals have similar powers. In some cases the power of thermo-luminescing can be revivified by fresh exposure to light, or by stimulation by an electric spark or by kathode rays. Wiedemann has found artificial substances that are thermo-luminescent, and in particular a preparation of sulphate of calcium having intermingled as a "solid solution" a small percentage of sulphate of manganese.

Here is a large specimen of quartz crystal weighing nearly a hundred pounds. One of its faces is almost flat. Taking a smaller crystal of quartz in my hand I rub it to and fro upon the larger crystal. You can all see the brilliant flashes that are emitted in the operation.

Reserving for discussion in my final lecture the use of electric discharges to produce luminescence, we will return to the properties of the ultra-violet light. One effect which they possess above all other kinds of light is that of producing diselectrification of electrified bodies, a phenomenon discovered by the late Professor Hertz. But there is this peculiar limitation. If the electrified surface is that of a metal surrounded by air, then when ultra-violet light falls upon it it will produce diselectrification if the surface is negatively electrified, but not if the electrification is of positive sign.[1]

The fundamental point is easily shown. Here is an electroscope made with two leaves of aluminium mounted on either side of a central blade of aluminium, and enclosed (Fig. 104) in a thin glass jar. To the top of the stem is affixed a disk of sheet zinc which has been freshly cleaned with a little sodium-amalgam. It is bent back at 45°, at which incidence the results are most favourable. I hold near the zinc disk a rubbed glass rod, and touch the disk while

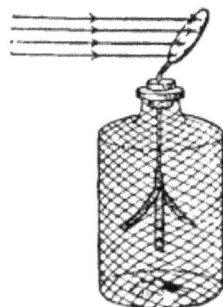

FIG. 104

---

[1] I have, however, found that a surface of peroxide of lead surrounded by an atmosphere of hydrogen is diselectrified if the electrification is positive.

it is under the influence of the positive charge of the glass. The electroscope thus acquires by influence a negative electrification, and the aluminium leaves stand out at a sharp angle. Now throwing a beam of ultra-violet light upon the disk, the leaves are seen to collapse rapidly. If the electroscope is positively electrified, the leaves do not fall down when the beam of ultra-violet light is directed upon the disk. Even the longer waves of visible light are active on a clean surface of sodium or potassium. The different kinds of light-waves have different photo-electric powers as well as different photo-chemical powers.

At the beginning of this lecture I dwelt upon the photographic actions of light-waves, and I return now to this topic in order to speak of the problem which has of late aroused such keen interest amongst scientific photographers, namely, the photography of colours. Many have been the attempts to produce true photographs of things in their natural colours. All hope of this was vain so long as photographers worked with chemically prepared plates that were more sensitive to the invisible light than to the visible kinds. Further, in the old collodion processes the greater sensitiveness of the chemicals to blue and violet waves, and their relative insensitiveness to orange and red light, caused all photographs to represent coloured objects untruly in their relative luminosity. It was an old complaint that blues photographed like white, and reds came out like black. The first steps towards remedying this arose in the discoveries of Vogel and of Abney that by staining the film or by giving to it in its preparation as an emulsion a fine granulation, its sensitiveness toward the longer visible waves might be increased. Thus were introduced the orthochromatic plates which gave as photographs a more accurate representation in black and white of the relative luminosities of objects; the ideal orthochromatic plate being one which should have the same relative sensitiveness toward the light of each part of the visible spectrum as our eyes have. Even before these discoveries the theory of the trichroic method of reproducing colour by photography had been enunciated by Clerk Maxwell.[1]   In the theory of colour-vision originated by Thomas Young, all colour-sensations are referred to three simple or primary colour-sensations, and it can be shown that no more than three[2] are needed to account for the various phenomena of colour-vision. These three primary sensations are the sensation of *red*, the sensation of *green* (a full green inclining to yellowish-green), and the sensation of *blue-violet* (a violet inclining toward blue). A red light stimulates but one of these sensations in the nerves of the eye. A yellow light stimulates two, namely, red and green, and is not therefore itself a primary sensation. Now if we could take three photographs of an object, each photograph corresponding only to the light of each primary sort, and if we could then illuminate each photograph with its own kind of light and superpose them, we ought to get a reproduction of the natural colours of the object. That, briefly, is the three-colour method.

[1] Discourse at Royal Institution, 17th May 1861.

[2] The reader should consult Captain Abney's treatise on Colour-vision.

The true photography of colours was only discovered a year or two ago by Professor Lippmann, whose exceedingly precious and beautiful results are individual pictures, incapable of being multiplied or reproduced. By placing at the back of the transparent sensitive film a mirror of mercury, each train of waves is reflected back during the exposure; and where the reflected waves meet the advancing waves of the train they set up stationary nodes that are spaced out through the thickness of the film at distances apart corresponding to the exact wave-lengths of the various lights. At these nodes the chemical action takes place, and produces a permanent picture which, when viewed by reflected light, shows all the natural colours of the object that has been photographed. I am, by the kindness of my colleague, Professor Meldola, able to show here, and to project on the screen, a Lippmann photograph of the spectrum in which all the colours show in their natural tints. More recently Professor Lippmann has shown in this theatre the remarkable colour-pictures which he has produced of landscapes, still-life subjects, and even of the human figure.[1]

Returning to the three-colour method of registering and reproducing by photography the natural colours of

[1] Since the delivery of these lectures two new processes of colour-photography have been announced. In one of these by M. Chassagnes certain chemicals are said to be used to treat the photographic films, by virtue of which they become capable of absorbing pigments at those parts of the picture which have been impressed by light of the corresponding tint. Another process by Mr. Benetto produces coloured transparencies by direct photography.

objects, I am happy in conclusion to be able by the
kindness of Mr. Ives to show you the remarkable results
which he has attained with his photochromoscope.
Starting from Young's theory of the three primary
sensations, Mr. Ives sought to construct colour filters
which should transmit for each of the three primaries
all those waves of the spectrum which excite that
sensation, and in proportion to their power of exciting
that sensation in the eye.   Thus the filter for red should
transmit not simply red light, but should transmit all
those waves of whatever colour that are competent to
excite the red sensation, but transmit them only in
proportion as they are competent to excite the red
sensation.   To select the proper tints as colour filters
is a matter of no small skill and experience.   Through
three such screens—one for red, one for green, one for
blue-violet—three photographs (negatives) are taken
simultaneously side by side upon a single orthochromatic plate.   From these three negatives (which are of
course colourless themselves) three positives are printed.
These also are colourless, but they show differences
according to the colours of the different parts of the
object photographed.   Fig. 105 is a triple chromogram
of a butterfly.   Its wings beside, having definite patches
of red and white on a black background, have all over
them a beautiful sheen of brilliant blue.   The upper-
most image of the three is that which is to be placed in
the blue-violet light.   The second figure is that for the
green light, while the lowest is for the red light.   Those
parts which are to show as white, when combined, are
white in all three images.   Each image is itself, like the

printed out, colourless; a mere black and white tran-
script on glass.

Now let these three colourless pictures be placed in an
instrument so arranged that blue-violet light falls through

Fig. 105.

1. To be illuminated by **Blue-Violet** Light.
2. To be illuminated by **Green** Light.   3. To be illuminated by **Red** Light.

the first, green through the second, and red through
the third of these separate photographs, and let them
then be recombined by suitable mirrors so that the eye
shall view them simultaneously, the primary colours

will recombine and give the object in all the glory of the natural tints.

The instrument (Photochromoscope or Kromskop) which Mr. Ives has designed for recombining these triple photographs stands upon the table. Fig. 106 gives a diagram [1] of its construction. Mr. Ives has also brought a lantern photochromoscope, by means of which he will now project on the screen a few of those beautiful photographs. The lantern itself has three nozzles, through which the

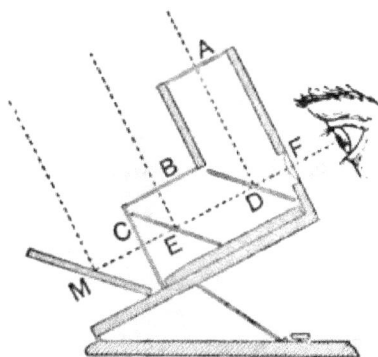

FIG. 106.

red, the green, and the blue-violet pictures are separately projected on the screen, and by their overlapping give the colour-combinations. He first shows us separately the three-coloured disks or circles of light, red, green, and blue-violet. Then he moves the nozzles so that

[1] A, B, and C are red, blue-violet, and green glasses against which the three corresponding transparent photograms are respectively placed. Two of the pictures at A and B are illuminated directly by light from above, the third C is illuminated by an oblique reflector. The red picture is viewed by rays which are reflected at the front surface of D an oblique transparent glass sheet. The blue-violet picture at B sends its light down upon another oblique transparent sheet at E, which reflects it through the sheet D to the eye. The green picture at C is viewed through both the transparent reflectors D and E. The lens F collects the rays for the eye, which thus views the three pictures as if they were superposed and at equal distance away. The instrument is made binocular, so the eyes see as it were a single image in its natural colours, and in solid relief.

the three disks partially overlap as in Fig. 107. Where
red and green mix they give us yellow; where green

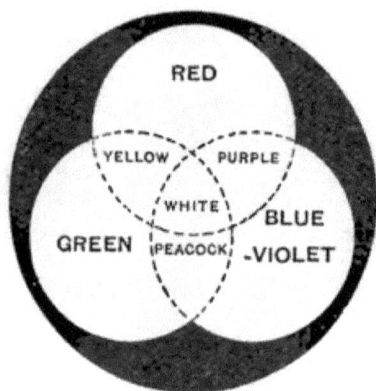

FIG. 107.

and blue - violet overlap
they give us a peacock
blue; where blue - violet
and red overlap they give
us purple; where all three
overlap at the centre they
give us white. Note how
the tint produced by the
overlap of two gives us the
complementary to the
third. Thus the yellow
is the complementary to
the blue-violet; the peacock is complementary to the
red; and the purple is of a tint complementary to the
green.

Those three photographs of the butterfly are now
introduced into the chromoscope-lantern, and are
brought to accurate superposition. The blue shimmer
on the insect's wings is shown with marvellous fidelity.
No painter could hope to produce by pigments such a
natural picture. Here is a photograph of a basket of fruit.
Note the yellow lemon. On examining the separate
colour-pictures one sees that this yellow is made up of the
red and green lights mixed. Here is a gorgeous bouquet
of flowers. The colours are superb. Here is a cigar-
box showing the natural brown tint of the wood; and
beside it a piece of *cloisonné* enamel, with all the delicate
shades of dull tints in their due relations. And lastly,
here is a box of sweetmeats, so naturally photographed

that one feels them to be really edible. After that we
need no further proof that by the proper selection of
the primary tints the dream has at last been realised of
registering and reproducing by photography the colours
of natural objects.

TABLE OF WAVE-LENGTHS AND FREQUENCIES

Name of Line.	Element.	Wave-length.		Frequency (billions per second).
		Micro-centimetres.	Millionths of inch.	
(Rubens and Nichols' longest waves)		2400	944	$12 \cdot 5 \, (\times 10^{12})$
(Langley's longest waves)		1500	592	20
(Paschen's longest waves)		945	370	31·7
Ψ_2 Ψ_1	...	270	106·24	111
Φ_2	...	124	48·73	242
Φ_1	...	120	47·25	250
Y	...	·004	35·36	333·7
	...	89·865	35·35	334·0
X_4	...	88·061	34·64	340·8
X_3	...	86·614	34·1	346·2
X_2	...	85·4	33·63	351·3
X_1	...	8 ·7	33·44	353·3
Z	...	82 ·4	32·34	364·5
A	O	75·94	29·28	395·2
B	O	68·674	27·03	436·5
C	H	65·630	25·83	457·2
D_1	N_2	58·961	23·21	508·8
D_2	N_2	58·902	23·18	509·1
D_3	He	58·760	23·13	510·5
E_1	Fe	52·705	20·78	569·2
	Ca	52·704	20·78	569·2
E_2	Fe	52·697	20·74	569·3
b_1	Mg	51·838	20·40	578·9
b_2	Mg	51·729	20·36	580·0
b_3	Fe	51·692	20·351	580·4
	Fe	51·691	20·350	580·4

TABLE OF WAVE-LENGTHS AND FREQUENCIES—*Continued*

| Name of Line. | Element. | Wave-length. | | Frequency (billions per second). |
		Micro-cen-timetres.	Millionths of inch.	
b_4 {	Fe	51·677	20·306	580·5
	Mg	51·675	20·305	580·5
F	H	48·615	19·14	617·1
G {	Fe	43·081	16·96	696·3
	Ca	43·079	16·95	696·4
h	H	41·018	16·17	731·3
H	Ca	39·686	15·63	756·0
K	Ca	39·338	15·48	762·7
L	Fe	38·206	15·04	785·1
M {	Fe	37·278	14·676	804·6
	Fe	37·271	14·673	804·9
N	Fe	35·813	14·09	837·7
O	Fe	34·411	13·55	871·8
P	Fe	33·613	13·23	892·6
Q	Fe	32·869	12·94	912·6
R {	Ca	31·814	12·52	942·9
	Ca	31·794	12·51	943·5
r	Fe	31·446	12·38	954·1
S_1 {	Fe	31·008	12·207	967·4
S_2 {	Fe	·004	12·206	967·6
	Fe	31·001	12·205	967·7
s	Fe	30·477	11·99	984·5
T {	Fe	30·212	11·894	993·0
	Fe	30·207	11·892	993·3
t	··	29·945	11·79	1002·0
U		29·480	11·60	1017·6
(Miller's limit, photog. phic)		20·2	7·95	1485·1
(Stokes' limit, fluorescent)		18·5	7·28	1621·6
(Schumann's highest frequency)		10	3·93	3000

LECTURE V

THE INVISIBLE SPECTRUM (INFRA-RED PART)

How to sift out the invisible infra-red light from the visible light—
Experiments on the absorption and transmission of invisible
infra-red light—It is cut off by transparent glass, but transmitted by opaque ebonite—Use of radiometer—Use of thermopile and bolometer—"Heat-indicating" paint—Experiments
on the reflexion, refraction and polarisation of invisible infra-
red light—Discovery by Hertz of propagation of electric waves
—Hertzian waves are really gigantic waves of invisible
light—Experiments on the properties of Hertzian waves ; their
reflexion, refraction and polarisation—Inference that all
light-waves, visible and invisible, are really electric waves of
different sizes.

To-day we deal with those waves of invisible light
which lie beyond the red end of the spectrum. They
are invisible to us because their wave-lengths are longer
than any to which the nerve-structures of our eyes are
sensitive ; or, to put it in the inverse way, because their
vibrations are of a frequency lower than any within our
range of optical perception.

Just as the ultra - violet waves have a shorter
wave-length and a higher frequency than the visible
waves, and have to be detected by their chemical,
luminescent, and diselectrifying effects, so the infra-red

waves of larger wave-length and lower frequency have to be detected and investigated by other physical effects than that of sight. The chief physical effect produced by these long infra-red waves is that of *warming* the things upon which they fall. For this reason they are sometimes called the *calorific* waves ; and the invisible light of this kind is sometimes [1] called "radiant heat."

But if, as I shall have to show you, this so-called radiant heat possesses (save in respect of visibility) all the physical properties of light ; if we can reflect it, and refract it, disperse it, diffract it, and polarise it, then we are logically compelled to admit that it is really a kind of light.

In brief, the spectrum extends both ways beyond the visible part ; beyond the violet are the *chemical* waves, and below the red are the *heat* waves.

If you find, as every one finds, that the light from the sun or from a flame warms the things on which it shines, it is natural to ask which of all the waves mixed up together in the beam give the warmth. To answer that question let us have recourse to the test of a carefully considered experiment. First let us spread out the rays into a spectrum, and then explore which part has the greatest warming effect.

The first explorations of the spectrum were made by putting into the different parts of the spectrum the bulb (blackened) of a thermometer. This showed

---

[1] Another term used by some writers is "the radiation." This use of the term is to be deprecated ; for the word radiation ought not to be used in two senses. If it is rightly used to mean the *act of radiating*, then some other term ought to be used to denote that which is radiated, namely, the waves.

that the heating effect is mainly at and beyond the red end of the spectrum.

The spectrum which is once again thrown on the screen (Fig. 108) is produced as on previous occasions by employing in the lantern an electric arc-lamp, in front of which is placed a slit, a lens to focus an image of the slit, and a prism to disperse the mixed waves into their proper places according to their wave-length.

Our spectrum to-day is neither so brilliant nor so extended as you have seen it on former occasions; the cause for this circumstance being that (for reasons you will presently appreciate) we are obliged to abandon the use of glass lenses and glass prisms, and substitute lens and prism of rock-salt. This material is less refractive and less dispersive, hence the narrowness of the rainbow-coloured band.

And now we have to make good by experiment the proposition that I have advanced that the heating effect is due to the longest waves—those at the red end and beyond the red end of the spectrum.

But as an ordinary thermometer would not be convenient I adopt another method, using instead a sort of electric thermometer—the *thermopile*. If you want to know all about this instrument, you must refer to treatises on electricity. All I need say now about it is that it is an apparatus, Fig. 109, which is exceedingly sensitive to heat, and which, when the face of it is warmed, generates an electric current. The electric current is led into a galvanometer which reflects a spot of light upon the scale against the wall. So, you may take it that that spot of light will indicate by its position

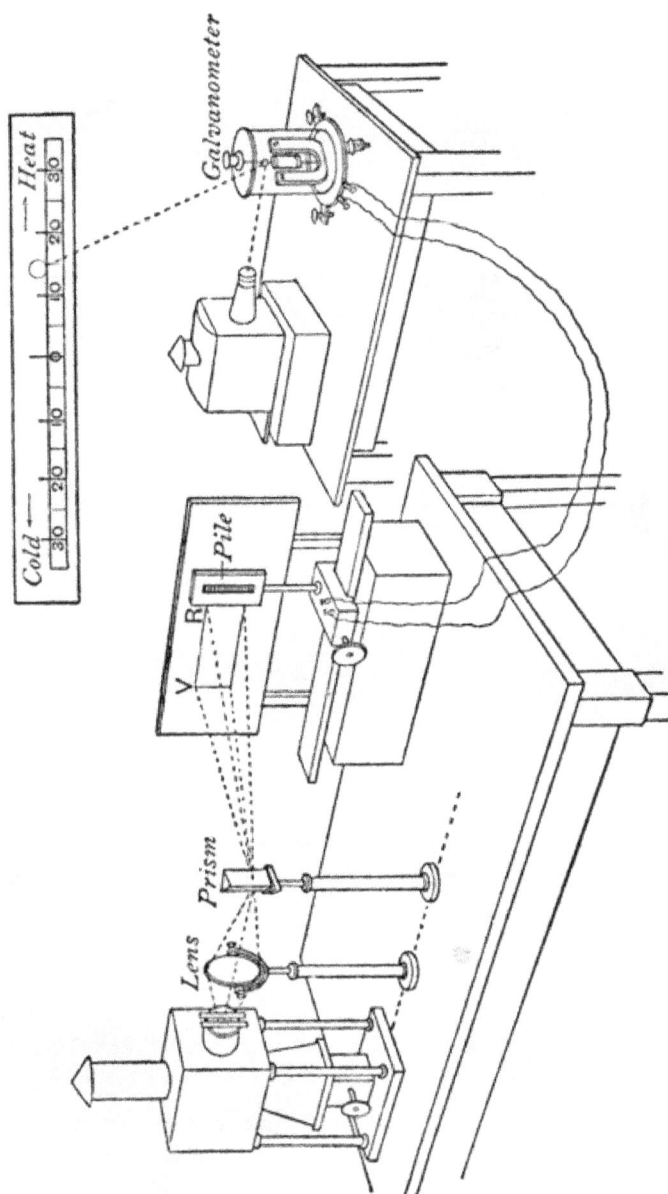

FIG. 108.

whether the face of the thermopile is warmer or colder

than the air of the room. If it is warmer the spot will move to the right ; if colder, to the left.

The spectrum now falls upon a small brass screen with a slit in it, behind which is the thermopile ; and at present the part of the spectrum that enters through the slit and falls on the face of the pile is the ultra-violet part. The spot of light is still at zero, showing that the ultra-violet light does not appreciably warm the face of

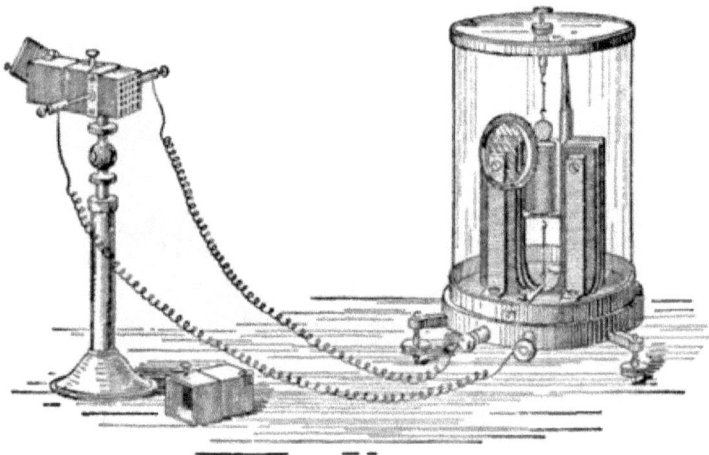

Fig. 109.

the pile. I now explore the spectrum by pushing the thermopile gently along. The slit now lies in the violet —yet there is no heating effect. The blue, the peacock, the green, and the yellow are successively explored— yet the spot of light remains at the zero. These waves do not produce appreciable heating. Another move forward, and the orange waves enter the slit and fall on the face of the pile—the spot begins to move. The orange waves warm slightly. I push on into the red,

and the spot moves gently across about a quarter of
the scale. · Red waves heat more than orange ones.
Pushing on beyond the end of the visible red (Fig. 108)
the effect increases. At a point about as far beyond
the end of the red as the red is beyond the green of the
spectrum, the heating effect is much greater—the spot
flies across the scale. Clearly the waves which have the
greatest calorific power are those some little way in the
invisible infra-red region : or in other words the waves
that heat most are waves having a wave-length somewhat
greater than that of the largest waves of the visible
spectrum. Taking the size of the extreme red waves at
32 millionths of an inch, we may put down these more
powerful invisible waves as about 40 to 45 millionths
of an inch in length.

The invisible infra-red spectrum has often been
explored, and by many explorers. Langley, using a
different electric instrument of his own invention,
termed a *bolometer*, has succeeded in observing waves
whose length was 592 millionths of an inch, or which
have a wave-length twenty times as great as those
of red light. Professor Rubens has independently
measured infra-red waves as large as 0·002400 centi-
metre, or about 944 millionths of an inch in length.
Hence, if set down in a scale of wave-lengths, the
infra-red spectrum stretches out to about fifty times
the extent of the visible spectrum. In the language
adopted for describing musical intervals, while the
range of visible light is about one octave (the extreme
violet having about double the frequency of the extreme red), the infra-red waves are known to go down

more than five octaves below, and the ultra-violet waves
ascend to about two octaves above the visible kinds of
waves (see the Table on pp. 190, 191).

Our thermopile has been placed at that part of the
spectrum, a little beyond the end of the visible red,
where we found the greatest heating effect. To
increase the effect somewhat, I will open the slit a
little, thus permitting a larger amount of these longer
waves to fall upon the face of the instrument. Having
thus adjusted our arrangements so as to be sensitive to
the heat-waves, we will try an experiment or two to find
whether these longer waves which produce the heating
effect are able to penetrate through the various materials
which we have tried for ordinary light. In the first
place, take a piece of window-glass, and try whether the
heat-waves will pass through it. On interposing it in
front of the slit we notice, by the indication given by
the galvanometer and thermopile, that though it cuts
off much of the heat it does not cut it off entirely.
Substituting a piece of red glass, we find that it also
cuts off some of the effect, but a blue glass cuts it off
much more. Now I take a piece of flint glass, which
contains lead : you note that it cuts off the waves much
more. Here is a slice of quartz crystal ; it does not cut
off the effect as much as the glass did. Again, here is
a slice of calc-spar of the same thickness. It cuts off
the heat-waves more completely than any of the materials I have yet tried. Lastly, here is a slice, also of
the same thickness, of rock-salt ; that is to say, a slice
of a big crystal of common salt sawn off and polished.
The rock-salt hardly cuts off the heat-waves at all.

Here then are four substances—glass, quartz, calc-spar, and rock-salt—all transparent alike to ordinary light. Quartz, as we saw in the last lecture, is exceedingly transparent to the ultra-violet kind of invisible light ; that is to say, to the shortest waves. But to-day we prove that rock-salt is the one that is most transparent to the infra-red kind of invisible light. Naturally, seeing that this fact was discovered half a century ago, we now apply the discovery in the construction of our apparatus. The lenses and the prism I am using to-day for these experiments are made neither of glass nor of quartz, but of rock-salt. And as rock-salt possesses a very poor dispersive power for the visible kinds of waves, it produces, as you have seen, but a poor spectrum of colours as compared with the spectrum that would be produced by the use of a prism of glass or quartz of the same size.

We might have used as an exploring apparatus, instead of our thermopile, another instrument, the *radiometer* (Fig. 110). Rather more than twenty years ago the celebrated chemist Crookes made the discovery that when light falls on movable things, such as the vanes of a lightly-poised mill, in a glass bulb from which the air has been mostly exhausted, so as to leave a fairly perfect

FIG. 110.

vacuum, the vanes of the mill are driven round. Apparently the blackened vane of the mill tends to retreat from the light. Why, we must presently consider. My present point is that it is possible to use this apparent repulsion to measure the intensity of the radiation that falls upon the instrument. Place the radiometer in one part of the spectrum; it turns round slowly. Move it on into the red end; it spins more quickly. But the effect is found to depend not merely upon the kind of waves, but also to some extent upon the nature of the surface of the vanes, and upon the degree of vacuum in the bulb. Some radiometers revolve most rapidly in the bright part of the spectrum to which our eyes are sensitive.

Whether we explore the spectrum with a thermometer, as Sir William Herschel did, or with a thermopile, or a bolometer, or a radiometer, we find that it consists of waves spread out in different directions, and that the different waves have different heating powers. And yet all these different waves, with their different powers, are emitted at one and the same time from the same source. If the thing that is heated is insufficiently heated it will not shine—that we all know. But even if not hot enough to shine visibly it will still emit some invisible waves. When you begin to warm a substance, at first it gives out only a few waves of very long wave-length. As you heat it more it gives out more of these heat-waves, and along with these heat-waves it also gives out some visible waves of shorter wave-length. If heated still hotter, so as to be white-hot, it gives out not only heat-waves of all sorts, but

visible·waves from red to violet, and also with ultra-
violet waves, all mixed up together.

In the next experiment we will employ a dark source
of waves. In short, we will test the waves that are
emitted from a vessel of hot water. Here is a beaker-
glass which I fill with boiling water from the kettle.
You would not see that in the dark, would you? In a
perfectly dark room it would not give out any of those
waves to which your eyes are sensitive. But if you were
to hold your hand a few inches away from it you would
feel a gentle warmth radiating from it. You can feel
with the nerves of your hand that which the nerves of
your eyes do not perceive, namely, the long waves or
calorific radiations. But these long waves warm all
things on which they fall. They do not, however, warm
them all equally. The fact can be established in many
ways. Black and dark substances absorb the waves
that fall upon them. Bright and shiny bodies reflect
most of the waves. What becomes of the waves that
fall on black and dark bodies? Their energy is not lost;
it is transmuted into sensible heat. Instead of wave-
motions in the free space, we have molecular vibrations
in the substance. The bright surface, such as polished
metal, upon which the waves may fall, is not warmed
by them, for the waves as they meet the surface are
not broken up, but simply start off again in some new
direction. Whenever waves break on a surface, and
are destroyed or absorbed in the action of breaking, the
result is heat. The so-called heat-waves, or infra-red
waves, are not themselves hot. They do not heat the
medium through which they travel as waves. But they

are readily absorbed by the things they fall upon, and, being absorbed, they warm that on which they fall. A black surface is one which absorbs both the invisible and the visible waves. It heats more readily than a white or a bright surface.

Now here is a peculiar thermometer (Fig. 111), having two bulbs, full of air, joined together by a bent

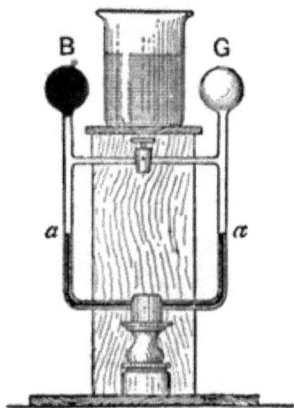

FIG. 111.

tube, containing a little coloured liquid to serve as an index: it stands up to the height marked *a*. If I put my hand on either bulb, and so warm the air inside it, the expansion of the air will depress the liquid in the tube below the bulb that is warmed. Were both warmed equally the liquid would not move. So this apparatus will indicate a difference of temperature, and is therefore called a differential thermometer. Next, note that one of the bulbs, B, has been painted dull black, the other, G, has been gilt with gold-leaf. I am going to put the beaker of hot water exactly in the middle between the two bulbs, where it can radiate equally to both of them. The gilt bulb, having a bright surface, reflects the waves, and is scarcely warmed at all by them. But the black bulb, having a more absorptive surface, will be warmed more than the bright gilt one; and you will see the indicating liquid fall in the tube below B and rise in the tube below G. Had we employed a beaker half blackened and half gilt, we could have readily demonstrated another point, namely, that a hot black surface radiates out the heat-waves more readily than a hot bright surface docs.

Now let us pass on to another experiment in which we again employ the thermopile.   Here is a thermopile connected by wires to the galvanometer, with its reflected spot of light on the wall.   It is arranged so that if the face of the thermopile is warmed the spot will move to the right and indicate to you the circumstance. In front of the conical mouth-piece of the thermopile

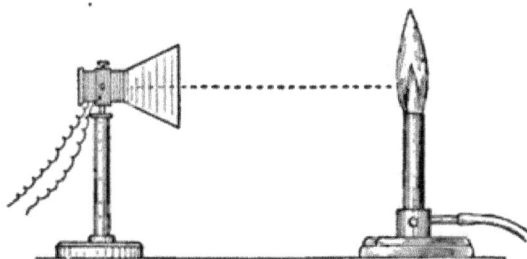

FIG. 112.

stands an ordinary Bunsen burner, which, as you know, is a gas jet having openings at the foot to let atmospheric air mix with the gas.   It gives a smokeless blue flame very different from the ordinary bright flame of gas. Though very hot, this flame radiates out but little light, neither does it, as a matter of fact, radiate off much heat. True, a column of hot air ascends from it straight up. But that is not what I am thinking of.   The question is, Is it sending out heat sideways?   Well, we can try. Opening the metallic shutter that closes the mouth-piece of the thermopile I let the light and heat, such as they are, radiate from the flame upon the face of the pile. At once the spot of light on the wall moves off to the right, showing that there are at any rate some waves present that can heat the pile. If I interpose for a moment a sheet of glass between the burner and the thermopile the spot of light comes back almost to its zero, showing that glass screens off nearly all the waves. I remove the screen, and the spot goes back to the right, showing that the heating effect has recommenced. Now comes the particular point of the experiment. If I stop up the holes at the foot of the burner where the air has been entering, the flame will at once burn brightly as an ordinary gas flame. The combustion will, as a matter of fact, be less perfect, for the flame will be sooty, and the total amount of heat produced in a given time will be less, because of the imperfection of the combustion. But also because of the imperfection of the combustion there are innumerable solid particles formed in the flame which get brilliantly heated, and emit light. They are better radiators of waves than the gaseous particles of the pale blue flame, and they radiate long waves better. To make the flame shine thus, I have but to stop up the air-holes with my finger and thumb; and instantly the spot of light on the wall rushes to the right, even beyond the end of the scale, proving that the bright flame radiates more heat-waves to the pile. I take away my fingers, air is readmitted, the flame relapses to its former pale state, and the spot of light settles back to its former position. Every time I let the flame burn brightly it radiates more waves sideways.

You may use a radiometer instead of a thermopile to demonstrate the facts. The vanes of the mill turn fast when the flame is bright, and more slowly when, by admitting air to the flame, you improve the combustion.

The next experiment I have to show you is with the same thermopile, only, instead of shining upon it with a flame, I will put a lump of ice in front of it. The spot of light on the wall now retreats, right beyond the zero mark, to the left, indicating that the face of the thermopile has been chilled. Perhaps you will say that this proves that the ice is radiating out cold. It may seem so. But that which is really occurring is this. "Cold" is a relative term meaning really "less hot." All things that are not in that unattainable state of absolute zero of temperature are more or less hot; hot things more, cold things less. And everything tends to radiate its heat away—the hotter it is the greater its tendency. Ice is less hot than the other things in this room. The ice is colder than the thermopile. The thermopile itself is radiating out heat, some of which goes to the ice. The ice is also, though to a lesser degree, radiating out heat. Here then we have two things, a thermopile which is warm, and ice which is colder, radiating to one another unequally. The result of this unequal exchange is that the thermopile parts with more heat than the ice parts with, and therefore is cooled. But the effect is the same as if the cold were being radiated.

Now let us go to an experiment that I believe took its origin nearly a century ago in this Royal Institution. In the Royal Institution the emission of heat and the properties of heat-waves have ever been favourite topics of study. The founder of the Royal Institution, Count

Rumford, himself originated many experiments on the
radiation of heat. It was he who discovered that heat
could be radiated across a vacuum. Sir Humphry
Davy, while Professor here, showed a most beautiful
experiment in which heat-waves were reflected from one
point of space to another by means of two paraboloidal
mirrors of silvered metal. That experiment I propose
to repeat, using the two mirrors which are believed to
be the actual pair used by Sir Humphry, and often
since used by Professor Tyndall.

One of the two curved mirrors is hung mouth-
downwards at a height of some fifteen feet above the
lecture-table. The other stands
mouth-upwards on the table
exactly beneath the first. The
upper mirror is lowered for a
moment. A red-hot iron ball
is slung by a hook in the focus
of the mirror, and it is hoisted
up again into its position above
the table. The ball being at
the focus, the mirror collects
the diverging waves and reflects
them straight down in a parallel
beam. It is a beam of invisible
infra-red waves, accompanied
by a few waves of visible red.
This beam falls upon the second

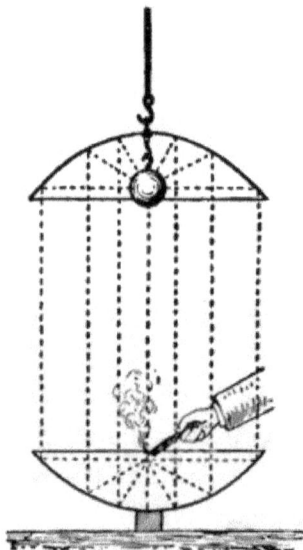

FIG. 113.

mirror (Fig. 113), which once more collects them and
converges them to a focus, F. I put my hand at
the place toward which the waves converge; it is

intolerably hot. I hold in the focus a bit of black
paper, at once it smokes, and kindles into visible combustion. Other things can be lighted. Here is a
cigar. Holding it in the focus it absorbs enough waves
to warm it up; and the ascending wreath of smoke
proves that it has been kindled by the reflected and
concentrated waves.

Our experiment has proved that heat-waves can be
reflected. Our earlier experiments with the rock-salt
prism proved that they can be refracted. Let us con-
firm these points by other experiments.

The red-hot ball, fast fading into dulness as it parts
with its store of energy, has now been placed upon a
stand on the table. Taking up the thermopile, which
did such useful service just now, we will see what it can
tell us about that red-hot ball. The distance between
the two is some seven or eight feet. I have, however,
only to turn the conical mouth-piece of the thermopile
straight toward the ball, and at once the spot of light
on the wall indicates that the thermopile has received
some of the radiation. Waves that our eyes cannot see,
this thermopile can see if only we turn it so as to look
straight at the source of the waves. It acts as a kind
of eye that is sensitive, not to visible light, but to infra-
red waves of invisible light. I turn the aperture of the
pile on one side so that none of the heat-waves can
enter it; the spot of light on the wall settles down to
its zero. Then, taking up a simple piece of tin-plate
to serve as a mirror I reflect some of the heat-waves
from the iron ball back into the mouth of the thermopile. As soon as the mirror is set at the proper angle

the spot moves to the right, showing that we have reflected some of the heat-waves into the pile.

While we have a hot ball—best if we had one that was brightly red-hot—I can show some interesting experiments which turn upon the employment of certain heat-indicating paints.[1] Here is a specimen of heat-indicating paint of a scarlet colour that turns black when heated. Here is another of pale yellow tint which turns red even when quite gently warmed. Here is a paper screen mounted in a convenient frame. The front is painted over with yellow heat-indicating paint: the back has been blackened that it may the better absorb the heat-waves.

[1] These heat-indicating paints are double iodides of mercury with other metals. They were discovered nearly thirty years ago by Dr. Meusel. The scarlet paint that turns almost black at about 87° C. is the double iodide of mercury and copper. The yellow paint which turns red at about 45° C. is the double iodide of mercury and silver. To prepare the former, a solution of potassium iodide is added to a solution of copper sulphate until the precipitate is redissolved, when a concentrated solution of mercuric chloride is added precipitating the red double-iodide. To prepare the more sensitive yellow paint, add to a solution of silver nitrate a solution of potassium iodide until the precipitate (silver iodide) redissolves. To this solution add a concentrated solution of mercuric chloride until a *bright yellow* precipitate is formed. The precipitates are collected on filter paper, and should be washed with cold water. They may be mixed with very dilute gum-water to enable them to be used as paint. With these paints many interesting experiments can be performed in illustration of the propagation of heat by conduction and convexion as well as by radiation. One very simple experiment is worthy of mention. It is to show how hot water will float on cold water. A strip of paper painted with the yellow paint is pasted vertically against the outside of a tall glass beaker. This is half filled with cold water. A floating disk of wood is introduced to prevent undue agitation, and then the beaker is filled up by pouring in boiling water out of a kettle. The top half only of the strip of paper turns red.

Holding it a little way from the hot ball—an ordinary coal fire answers even better—I place my hand between the ball and the screen, against the back of it. In a few seconds the screen turns red all over except where it is protected by my hand, of which a shadow—a sort of heat-shadow—in yellow is temporarily photographed, or rather thermographed, upon the screen. As the screen cools it returns to its former yellow tint.

Here is another screen made of paper painted with scarlet heat-indicating paint. The back has been gilt all over, and then on the gilt surface a big letter S has been painted. I hold this with its gilt back to the hot ball ; and the gilt surface reflects away most of the heat-waves. Now you might suppose that the part where black paint has been put on over the top of the gold would be doubly protected against heat. But, no ! It absorbs the waves and grows warm ; and the heat being conducted through the gold film causes the scarlet paint on the front of the screen to turn black. The letter painted on the back is visible on the front of the screen.

Here is a variation upon one of Professor Tyndall's observations. You will find it recorded in his book [1] on heat how, on one occasion when a fire broke out in a street, the heat radiated across the street from the burning house, charred the window-frames and burned and blistered the paint on the sign-boards. But where the number of the house stood in gilt letters on the sign-board the mere film of metal had reflected away the waves, protecting paint and wood behind it from

[1] *Heat a Mode of Motion*, p. 263.

being charred. In illustration, here is a blackened sheet of paper upon which a triangle of gold - leaf has been pasted. The other surface of the paper has been coated with the scarlet heat - indicating paint. Exposing it to the radiation of the hot ball you see how the triangular space protected by the gold-leaf remains cool, while the rest absorbs heat, turning the scarlet to black.

You will probably admit that we have now plenty of proofs that these invisible heat-waves are really a kind of invisible light: that the difference is one of degree rather than of kind. Consider yet again the process of *incandescence* in which such waves are emitted. Heat a body, beginning by gently warming it. At first its particles vibrate but moderately; the waves they send out into the surrounding ether are few and of relatively great wave-length. As you warm the body more and more its particles vibrate more actively, they jostle together; it gives out more waves and waves of shorter length and higher frequency. There are still the long waves, in fact there are more long waves than before, but there are some shorter waves in addition. Heat it still hotter. The lower kinds of waves still continue to be emitted, nay, are emitted more copiously, but some waves of a still higher kind now accompany them. Here is a thin platinum wire stretched between two supports. By leading into it through a thicker copper wire an electric current I can heat it as little or as much as I choose. It is now warm, giving out a few dull waves. Increasing the current its temperature is raised, and now it gives out much more heat, and with the heat a few waves of the visible red

sort. Every solid body when heated shows red as its
first colour on heating. Never is the first glow[1] of a
blue or yellow hue. Increasing the temperature of the
wire it emits orange light as well as red, and looks therefore bright red. The next increase brings in yellow
along with orange and red : then green comes in to join
the yellow, orange, and red. So soon as the wire is
heated so hot as to give out all the different visible
kinds, so soon we call the state a white heat. But
no solid ever gets blue hot, because in all cases the
emission begins at the bottom of the spectrum with red,
the other colours chiming in until white is attained.
Nor is white[2] attained until a certain proportion of the
still higher ultra-violet waves are being also emitted. So
then it appears that the process by which visible light-
waves are emitted is only a continuation of the process
by which the invisible infra-red waves are emitted. What
further proofs do you require as to the essentially kindred
nature of the visible and invisible waves ? I have shown
you that these infra-red waves behave as visible light-

[1] Captain Abney has shown, however, that owing to the want of
sensitiveness of the eye for red light, and its greater sensitiveness for
green light, the tint of *minimum visible luminosity* of any hot body
or indeed of any feebly illuminated body in a perfectly dark room is
greenish. This is true even of a light seen through ruby glass if the
eye has been kept some time in darkness.

[2] The whitest known artificial light is that of the arc-lamp ; it is
the light of carbon incandescent at about 3500° C. This is a
temperature considerably lower than that of the sun's surface, which
emits a light having a relatively higher proportion of blue and violet
and of ultra-violet waves. In fact, when seen in full sunlight the
light of the arc-lamp is decidedly dull and reddish. No accurate
definition of any standard of *whiteness* has ever been given.

waves do in a number of respects. You have seen that
we can refract them with a lens, disperse them with a
prism, reflect them with a mirror, and absorb them with
a black surface. Further, they travel at the same rate
across space as the visible waves do. This we know
from that which happens at the time of a total solar
eclipse. At the moment when the sun's light ceases to
be visible, his heat ceases also to reach us. When the
light reappears the heat-waves are also restored. This
one fact proves these heat-waves to be simply light of
an invisible kind. But if you are not satisfied I will give
you yet one further proof. You shall see that they can
be polarised.

Here, as in my third lecture (Fig. 93, p. 132), stand a
pair of Nicol prisms, one to serve as polariser and the
other as analyser. The lantern sends its beams through
them. Receiving the visible light on a paper screen we
note that when the analyser is set with its principal
plane parallel to that of the polariser light is transmitted :
but on rotating the analyser through a right angle all
light is cut off. That is a purely optical experiment.
Now let me take my thermopile—which I described to
you as a sort of eye which is sensitive to the invisible
heat-waves—and put it in the place where the paper
screen was. At present the polariser and analyser are
crossed, giving the "dark field" (p. 119). No light falls
on the thermopile, nor any heat-waves, for, see, the spot
of light from the galvanometer that indicates the state of
the thermopile is at its zero point. Now I turn back
the analyser and restore the bright field. At once the
spot of light on the scale swings over to the right, telling

us that the polarised heat, as well as the polarised light, is coming through the analyser. Turn the analyser back again, the visible waves are cut off, and so are the invisible ones, for the spot of light has returned. Now let me clinch the proof by working entirely with invisible waves to the exclusion of visible ones. Here is a sheet of opaque hard black indiarubber, of ebonite in fact. No visible light will come through it. But yet, you observe, when I have thus filtered out the invisible waves, and stopped off the visible ones, still there come through the polariser some waves which can warm the thermopile, and which can be cut off by turning the analyser to the position at right angles.

This material ebonite is a most interesting one from the circumstance that it can thus act as a wave-filter transmitting only the longer waves. Wave-filters (or ray-filters) were extensively used by Professor Tyndall in his lectures on radiant heat: but I do not think he was acquainted with the properties of ebonite. Here is one of the Crookes radiometers (Fig. 110, p. 199). I place it in front of the lantern but screen it at first by a thick sheet of metal so that it is all but at rest. The diffused light in the room suffices to make it turn slowly. I substitute for the metal sheet a sheet of ebonite which is equally opaque to ordinary light. Yet the little vanes now run merrily round.

Tyndall's filter for heat-waves consisted of a cell containing a dark solution of iodine in bisulphide of carbon. Here is one of the cells, kept cool by an outer jacket in which cold water circulates. Behind the wall at the back of the theatre is a powerful electric arc-lamp, the beams of which pass into the theatre through an aperture. This beam I propose to concentrate by a rock-salt lens, bringing it to a focus, after it has passed through the cell that filters out the invisible waves and stops the visible ones. First we make the experiment without the cell. All the waves visible and invisible come to a focus. Holding at the focus a bit of black paper it smoulders and then takes fire. Now interpose the cell. The visible light is cut off; but holding the bit of paper in the invisible focus it again begins to smoulder and finally breaks into visible burning.

And now I have to pass to the most important recent discoveries—discoveries dating only from 1888—of some larger waves which are exactly like light-waves in the following respects: they can be reflected, refracted, absorbed, polarised, and diffracted. Yet they differ in the most striking way from any of the waves of light that we have hitherto considered. Their wave-lengths, instead of being measured by a few millionths of an inch, may be several inches, several yards, or even several hundreds of yards long. I refer to the *electric waves* predicted in 1864 by the late Professor Clerk Maxwell, and discovered experimentally in 1888 by the late Professor Hertz.

Hertz was occupied with researches upon electric sparks, which, under certain circumstances, were known to be oscillatory. That is to say, each spark might, under certain conditions, consist of a series of sparks flying backwards and forwards along the same path with great regularity and excessive rapidity. If, for instance, there were twenty successive oscillations, each lasting only one one hundred-millionth part of a second,[1] the whole series would only last one five-millionth part of a second, and would, of course, seem to the eye as simply an instantaneous spark. In working with these oscillatory sparks Hertz was led to investigate the disturbances which they set up in the surrounding medium, and

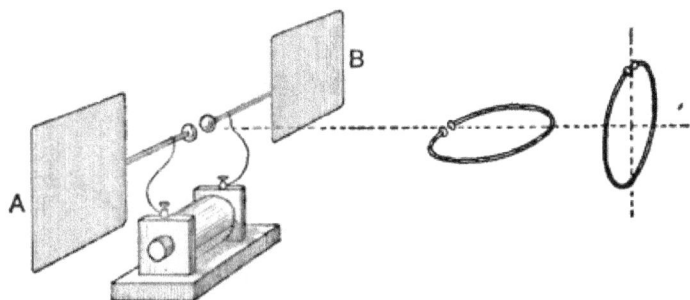

FIG. 114.

which are propagated as waves. To illustrate Hertz's work I must have recourse to a few diagrams. Fig. 114 illustrates the apparatus which is set up on the table. To produce the sparks we employ an induction-coil. The electric discharges produced by the coil are sent into the simple apparatus called by Hertzian *oscillator*. As you see it consists of two square sheets of metal, affixed upon two metal rods that nearly meet and are

[1] It may be useful to note that since the velocity of propagation of electric waves in air (or vacuum) is identical with that of light (186,400 miles per second, or 30,000,000,000 centimetres per second), the wave-length can be deduced from the frequency by the rule that the product of frequency and wave-length is equal to that velocity. In the above example, if the period is one one hundred-millionth of a second, the frequency is one hundred million a second; dividing 30,000,000,000 by 100,000,000 we get as the wave-length 300 centimetres, or about ten feet as the wave-length.

provided with two well-polished metal balls as terminals.
There is a small gap between which the sparks are seen
to pass.   But each such spark is really a series of oscillations; the electric discharge oscillating backwards and
forwards, not simply across the gap where you see the
spark, but from one end to the other of the apparatus.
Suppose the coil to make one of these metal wings (say
A in Fig. 114) positive, while the other wing (B in Fig.
114) is negative.   When the electric state has risen
sufficiently high, the air in the gap is pierced by a spark.

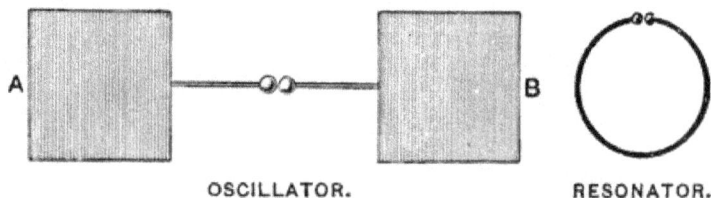

OSCILLATOR.   RESONATOR.

FIG. 115.

The charge rushes from A to B, and in so doing over-
charges B, making it positive, while leaving A negative.
At once the charge surges back again from B to A, and
again back to B, each oscillation lasting only about the
one hundred-millionth part of a second.   The frequency
of the oscillations depends on the size of the apparatus.
At every oscillation an electric wave is sent off from the
apparatus into the surrounding space, and is propagated
with the velocity of light.   The wave is propagated with
the greatest intensity in the directions at right angles to
the metal rods along which the electricity is oscillating,
and at right angles to the plane of the metal wings.
Fig. 115 gives a front view of the oscillator, and also of
the apparatus called the resonator used by Hertz for

detecting the waves.   The model on the table is made
of the same size as one of Hertz's smaller pieces of
apparatus, the two metal wings being each 40 centimetres square, and the distance between them 60
centimetres.   The wave-length of the waves emitted is
rather less than 300 centimetres, or nearly 10 feet.
The resonator or detector is a simple wire, bent into a
ring so that its two ends nearly meet.   Hertz demonstrated the fact that waves pass from the oscillator by
holding the resonator some distance away from it, and
observing the minute electric sparks which they set up
in the small gap between the ends of the wire.   But it is
necessary that the resonator ring should be of the proper
size, and that it should be held in the right position.
The size (in this example 70 centimetres diameter) should
be such that the natural period of oscillations of an electric current around the ring should agree with the period
of the waves emitted by the oscillator.   The position
should be such that as the waves from the oscillator
reach the resonator they set up secondary oscillations
in the ring.   If the resonator is set up vertically edge-
ways to the oscillator, no sparks are produced : the
waves simply stream past the resonator.   If, however,
the resonator is held horizontally, and in the base-line
shown in Fig. 114, sparks may be detected in the gap.
Hertz put at the far end of the room where he was
working a great sheet of metal to reflect back the waves,
and then went about to different positions in the room
exploring the space to find at what points sparks were
produced.   He found that when the waves are thus reflected back on themselves there are nodal points, just

as there are nodal points in sound-waves and in light-waves when reflected back. These nodal points were spaced out at distances apart exactly equal to half the wave-length, which thus could be precisely measured.

Before Hertz's time it was indeed known that there were oscillating sparks. Fig. 116 illustrates some experiments which I myself made [1] in the year 1876 on this subject. I had an induction coil connected to send sparks

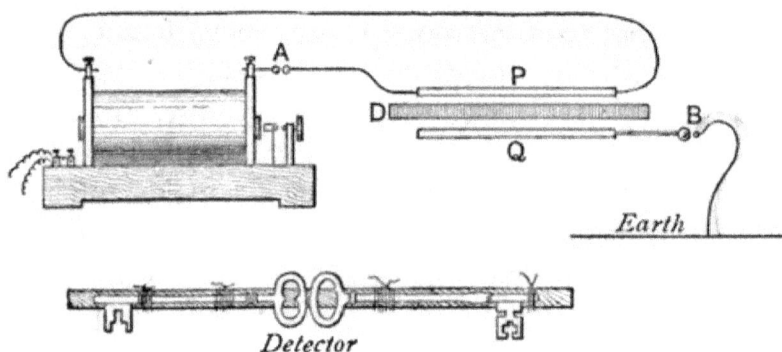

Fig. 116.

across a small air-gap, A, to a condenser made of a dielectric, D, between two metal plates, P and Q. I found that if there was this spark-gap in the circuit of the coil I could draw secondary sparks at B from the outer plate of the condenser; and by means of a small vacuum-tube and a rotating mirror I proved that these sparks were oscillatory in character. When these arrangements were made I was able to get sparks from insulated metal objects in the room. These sparks could be traced all about the room. I had but to hold a knife or pencil-case to the

[1] *Philosophical Magazine*, September 1876.

FIG. 117.—PROFESSOR HEINRICH HERTZ.

door-knob or other piece of metal to draw sparks. I
even did this: I took two door-keys and tied them on to
a piece of wood, so as almost to touch one another, and
with this detector I could get sparks while walking about
to different parts of the room. But it never dawned
upon me that these sparks were the evidence of electric
waves crossing the space. That was Hertz's discovery.
He did not go idly about the room noticing the sparks,
but explored the positions where the sparks were to be
detected, and holding his apparatus in the right position
to detect them.

A word more about the electric oscillations themselves. Each sudden discharge of the induction coil—and
to make them sudden
the discharge balls
must be well polished
— sets up a set of
oscillations, diagram-
matically represented
in Fig. 118 by the
upper curve, which die
away as time goes on.
A mechanical analogy
may be found in the

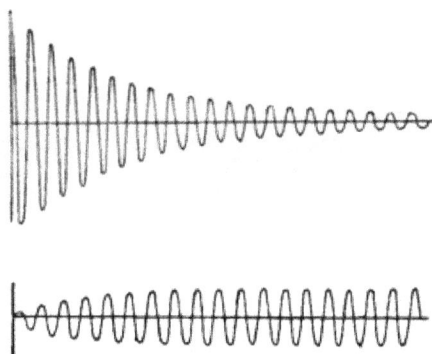

Fig. 118.

vibrations of a spring. Bend it on one side and suddenly release it. It flies backward and forward, the
motion dying out after a certain number of swings.
So trains of waves are set up, which also die away as
they travel across space. But suppose they fall upon a
proper resonator or detector, then they will set up, by
their timed impulses, a sympathetic electrical vibration

in that resonator, the oscillations thus set up beginning and increasing in strength as wave after wave arrives. This is represented graphically by the lower curve in Fig. 118. This corresponds, in fact, to the way in which the sound-waves from a tuning-fork, when they fall upon another tuning-fork, will set it into sympathetic vibration, provided it is tuned to the same note.

In Fig. 119 are represented the parabolic mirrors, each about 6 feet high, with which Hertz demonstrated the

FIG. 119.

reflexion of electric waves. At the focus of one of these mirrors there was placed vertically an oscillator, an arrangement to produce sparks vibrating up-and-down. The waves which resulted were, of course, waves of up-and-down motion—polarised in a vertical plane—which were reflected in a beam straight across the room to the second mirror, which collected them and reflected them to a focus upon a detector, which in this case was straight, not circular, with a small spark-gap at its middle, where the minute sparks could be detected.

Many forms of oscillator or vibrator have been used

by different experimenters to produce electric waves.
Some of these are shown in Fig. 120.   The first is one
of those used by Hertz himself.   Instead of flat wings of
metal he used in this case cylindrical metal conductors.
In another form, described as a dumb-bell oscillator,
there were two large metal balls.   In every case the
spark-gap was arranged between two small highly-
polished metal balls, midway along the length of the
oscillator.   The third shape is one devised by Professor

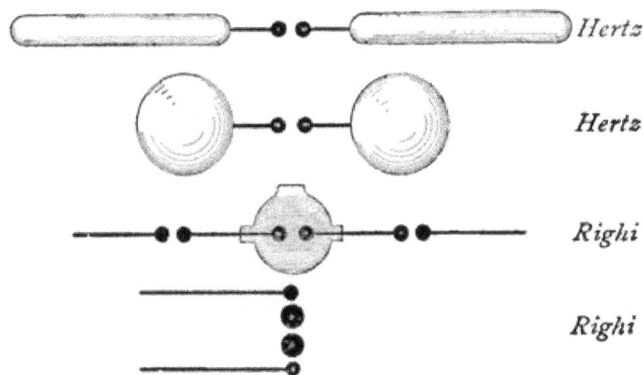

FIG. 120.

Righi for making short waves.   Here there are three
gaps, the central one being between two balls immersed
in an oil vessel to prevent premature discharges.   The
lowest form in Fig. 120 is also of Professor Righi's devis-
ing.   It represents his apparatus for producing exceedingly short waves—less, in fact, than an inch long—by
the oscillations set up between two spheres to which
sparks were communicated from two smaller terminal
balls outside them.

The next diagram (Fig. 121) depicts two forms of

oscillator used by Professor Oliver Lodge, of Liverpool. Here is a well-polished metal ball supported between two smaller balls that nearly touch it, one on each side. When a discharge is made through this central ball, an electric charge surges from side to side in it with great vigour, but the motion dies out after only about three or four such surgings, since it readily radiates its energy into space. It does not emit a long train of waves; the

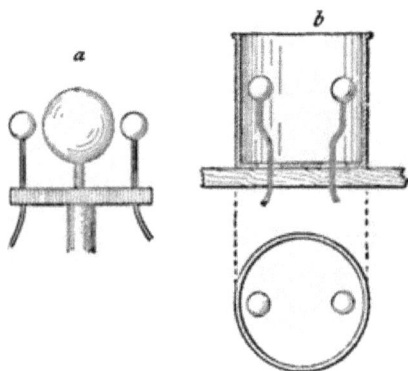

FIG. 121.

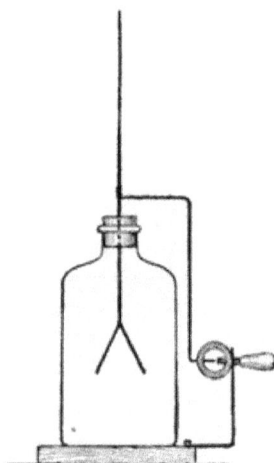

FIG. 122.

effect dying out after about $1\frac{1}{2}$ or 2 complete oscillations. The wave-length of the emitted waves is about $1\frac{1}{2}$ times the diameter of the ball. The other form, Fig. 121b, shows a metal cylinder which is sparked into at opposite ends of a diameter by two interior balls. This oscillator emits its energy less rapidly, and the oscillations last longer. It gives rise to a train of waves which are propagated chiefly straight out of the mouth of the cylinder. Fig. 122 depicts one of the simplest ways of detecting such

electric waves, and at the same time makes them evident
to a whole audience.   An ordinary gold-leaf electroscope
is provided with a by-pass of wire arranged with a minute
gap (adjustable by a screw) to break the continuity.   If
the gold leaves are carefully charged they will remain
diverging, because their electric potential has not been
raised sufficiently to cause a discharge across the gap.
But if now an electric wave from a Hertz oscillator—
especially from an oscillator set vertically to produce
vertically polarised waves—falls on the electroscope, it
sets up in the wire by-pass an electric surging that will
overleap the gap with a minute spark.   And, during the

FIG 123.

time that the spark bridges the gap the gold leaves discharge themselves and fall together.

A still more sensitive detector used by Lodge consists
of a bit of glass tube filled loosely with iron filings
(Fig. 123) and joined along with a weak voltaic cell in
circuit with a galvanometer.   Loose metal filings or
powders form a very bad and incoherent conductor:
hardly any current passes through them.   But let an
electric wave fall on the tube, instantly the filings become
—as discovered by M. Branly—an excellent conductor.
So there results a movement of the galvanometer, and of
the spot of light reflected by it, proving that an electric
wave has been detected by the tube.

Fig. 124 shows a set of the apparatus with which

Professor Lodge repeated and verified the observations of
Hertz as to the optical properties of these electric waves.
Hertz had reflected them with parabolic mirrors 6 feet
high, and refracted them with huge prisms of pitch. He
found they could penetrate through wooden floors and
stone walls.   He polarised them and diffracted them.
Lodge was able to repeat these results, but with an
apparatus of less heroic dimensions.   The sender con-

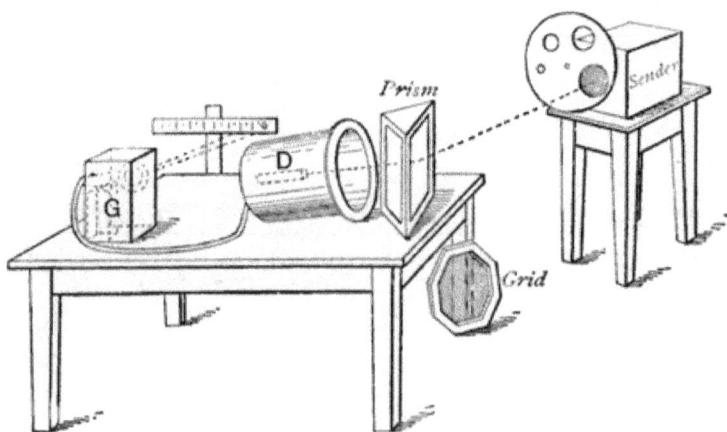

FIG. 124.

sists of an oscillator, like Fig. 121a, having a 5-inch ball
emitting 7-inch waves, enclosed in a copper box furnished
in front with a diaphragm perforated with apertures of
various sizes to moderate the radiation to any desired
degree.   As detector, D, there is used a tube full of
coarse iron filings put at the back of a copper hat, whose
open end is turned in any direction in which waves are
to be received.   Wires pass from the detector to the
galvanometer, G, and are enclosed in a metal tube to
shield them from stray radiations.   If the receiver is set

obliquely to the sender so that no waves from the sender
enter the receiver, the galvanometer will give no indica-
tions.   But if the waves from the sender are reflected
into the mouth of the receiver by holding in front, at the
proper angle, a sheet of metal, at once the detector is
affected, and the galvanometer reveals the fact.   Similarly, as shown in the diagram, a prism of paraffin wax
may be used to refract the electric waves into the mouth

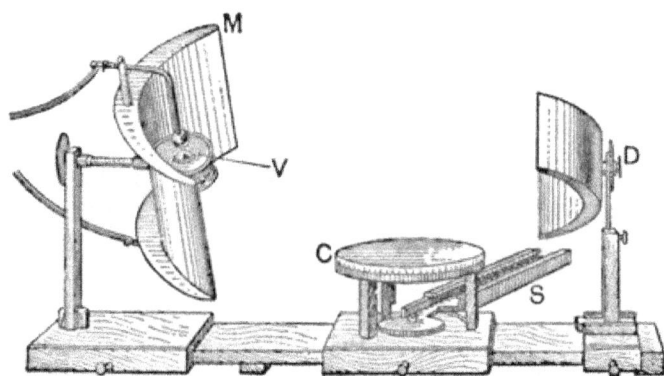

FIG. 125

of the receiver.   The picture also shows a grid of metal
wires which can be used to polarise the electric waves.

   More recent than the researches of Professor Lodge
are those of Professor Righi, whose apparatus is shown
in Fig. 125.   It consists of two parts, a sender and a
receiver.   The sender, on the left, consists of a small
oscillator (that shown at the bottom of Fig. 120), with
three spark-gaps, the central gap being capable of fine
adjustment.   In Fig. 125 this gap is between the ball
marked V, and one below it (not shown), enclosed in a
small leather capsule filled with vaseline.   This oscillator is set at the focus of a parabolic mirror, M, to reflect

the waves out straight to the right across the central table C. The detector, D, which also is furnished with a parabolic mirror, is an optical one. It is made of a film of silver upon a slip of glass, the film being divided in two across the middle with a diamond cut. Across this narrow gap minute sparks pass, and are viewed through an eye-piece at D. The apparatus is quite small enough to be put upon an ordinary table, and presents quite the appearance of a piece of optical apparatus. Upon the central table can be mounted reflectors, prisms, lenses, grids, or any other apparatus. With these devices Professor Righi has tracked down the optical properties of the electric waves varying from 8 inches down to 1 inch in length. He has demonstrated interference fringes by Fresnel's mirrors, and with the biprism, with thin plates, and by diffraction. He has verified the laws of refraction, reflexion, total reflexion, polarisation, and of elliptical and circular oscillations. He also investigated the transparency of media and the selective transparency of wood according to its grain, a property which makes it polarise the electric waves just as tourmaline polarises light. In short, he has completed the proofs that these waves possess all the known properties of ordinary light. Other workers have occupied themselves in the same field. We are shortly to hear a discourse here [1] by Professor J. Chunder Bose, of Calcutta,

[1] Given Friday, January 29, 1897. Professor Bose's oscillator is depicted in Fig. 126; it is made of a small ball of platinum between two smaller balls. Single sparks are given to this from a small induction-coil. A cylindrical lens of ebonite in front of this oscillator renders parallel the emitted waves. The complete apparatus is shown in Fig. 127. The oscillator or sender S enclosed

upon the polarisation of the electric wave as studied by him, with an exceedingly elegant apparatus producing still shorter waves.

But before I close I must show you at least some of in a metal tube projects from a box A, doubly cased in metal; which contains the induction-coil and battery.   The detector D consists of a number of small metal springs lightly pressing against one another, traversed by a current from a single cell C in the circuit of which is in-

FIG. 126.

cluded the galvanometer G.   The detector D is set up on a movable arm on an optical circle, so that the optical properties of the electric beam may be studied.   M is a plane mirror, N a curved mirror for studying the laws of reflexion, P is a prism of ebonite, T a special apparatus for observing the total reflexion

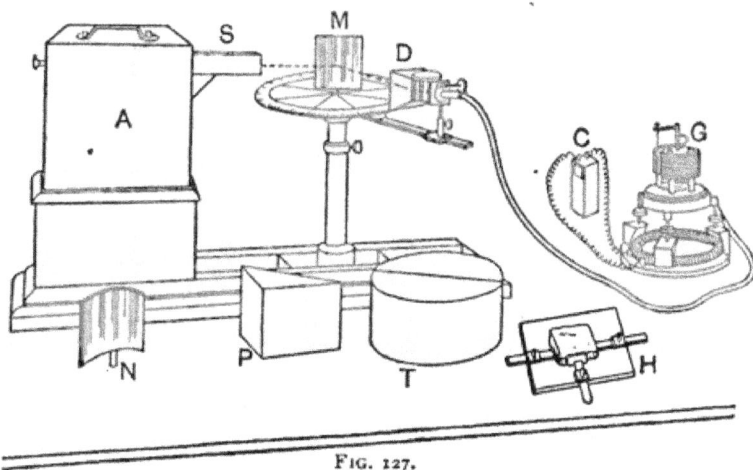

FIG. 127.

at films of liquid enclosed between two semi-cylinders of ebonite, H is a holder in which pieces of minerals can be clamped for observation.   Professor Bose has found that many crystalline and fibrous minerals, such as epidote and asbestos, polarise the electric

these effects in actual experiment. Here, in a metal box, is a small induction coil actuated by a couple of battery cells. The spark which this makes is carried to a small oscillator, closely resembling Lodge's (Fig. 121). It is, in fact, a short piece of polished platinum tube, 4 millimetres in diameter, between two small beads of platinum. It emits waves about $\frac{1}{2}$ inch long. I surround

it with a metal bonnet or tube to direct the waves straight out. The detector is simply a tube loosely filled with iron filings in circuit with a galvanometer, and a cell made with a bit of iron and a bit of copper dipping into salt water. It also is furnished with an outer metal tube to screen it from stray radiations. If I set the sender and receiver opposite one another, the receiver will respond whenever a spark is passed through the sender. After each such response, I have to give a gentle tap to the detector to shake up the filings and send the galvanometer index back to zero.

Now I set the receiver askew so that none of the waves from the sender shall get to the detector. Spark after spark is discharged across the oscillator, but there is no response. Then I hold a plate of metal as a mirror

beam, since (like the tourmaline for short waves) these materials absorb the electric vibrations in directions parallel to certain axes of their structure. Even an ordinary book possesses for these waves a polarising structure ; the waves that vibrate parallel to the leaves being absorbed more than those that vibrate in a direction transverse to them.

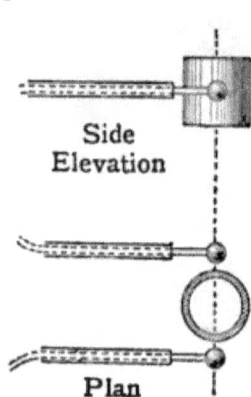

Side Elevation

Plan

Fig. 128.

to reflect the waves, or I interpose at the proper angle a
small prism of paraffin wax. At once the detector
responds, proving that the waves have been turned
round the corner, refracted by the prism.

If then it can be proved that these electric waves,
though invisible, can be reflected, refracted, polarised,
and absorbed, exactly as the visible waves of ordinary
light, have we not good reason to regard them as one
and the same phenomenon? By every test, in every
physical property, save only the accident that our eye is
not sensitive to them, they are nothing else than waves
of light. But if that is so, are we not entitled logically
to draw the converse inference that if light, ordinary
light, behaves in the same way and has all the same properties on the small scale as these electric waves on the
larger scale, then the little waves of ordinary light are
also electric waves? That, indeed, was the brilliant
speculation, the daring theory propounded in 1864 by
the late Professor Clerk Maxwell. Basing his ideas upon
the investigations pursued in this institution by Faraday,
who himself ventured first into this enchanted domain of
electro-optics, Maxwell predicted the properties of electric
waves in that famous memoir wherein he set forth the
doctrine that light consists of electric vibrations in space.
And the brilliant success of Hertz and those who have
followed him in demonstrating by experiment the optical
properties of these waves, is the abundant justification of
Maxwell's prediction.

APPENDIX TO LECTURE V

The Electromagnetic Theory of Light

DURING the first quarter of the present century the wave-theory of light successfully displaced the older corpuscular theory. Young, Fresnel, Arago, Biot, and Airy established the laws of physical optics upon an unimpregnable basis of undulatory theory, leaving Brewster the sole surviving exponent of the material nature of light. But none of those who thus contributed to establish the wave-theory of light could do much to elucidate the nature of that wave-motion itself. If light consist of waves they must be waves in or of something : that something being provisionally called the ether. But as to the nature of the ether itself, or as to the particular motions of it that were propagated as waves, scarce anything was to be learned save that the ether itself behaved rather like an incompressible liquid or solid of extreme tenuity but great rigidity, and that the waves were of the kind classed as transversal (see p. 108).

In 1845 Faraday discovered the singular fact that the magnet exercises a peculiar action on light; the plane of polarisation of a polarised beam being rotated when the beam passes along a magnetic field.

The existence of a relation between light and magnetism being thus established, Faraday proceeded to look for other relations, including the action of an electrostatic strain on polarised light, and the effect of reflecting polarised light at the polished pole of a magnet, neither of which, however, he succeeded in observing.

In 1846 he sent to the *Philosophical Magazine* some

"Thoughts on Ray Vibrations" in which he suggested that radiation of all kinds, including light, was a high species of vibration in the lines of force. "Suppose," he says, "two bodies A B, distant from each other and under mutual action, and therefore connected by lines of force, and let us fix our attention upon one resultant of force having an invariable direction as regards space ; if one of the bodies move in the least degree right or left, or if its power be shifted for a moment within the mass (neither of these cases being difficult to realise if A and B be either electric or magnetic bodies), then an effect equivalent to a lateral disturbance will take place in the resultant upon which we are fixing our attention ; for either it will increase in force whilst the neighbouring resultants are diminishing, or it will fall in force as they are increasing. . . . The propagation of light, and therefore probably of all radiant action, occupies *time;* and, that a vibration of the line of force should account for the phænomena of radiation, it is necessary that such vibration should occupy time also. . . . As to that condition of the lines of force which represents the assumed high elasticity of the æther, it cannot in this respect be deficient : the question here seems rather to be, whether the lines are sluggish enough in their action to render them equivalent to the æther in respect of the time known experimentally to be occupied in the transmission of radiant force."

In 1864 Clerk Maxwell, in a paper in the *Philosophical Transactions* on "A Dynamical Theory of the Electromagnetic Field," wrote :—"The conception of the propagation of transverse magnetic disturbances to the exclusion of normal ones is distinctly set forth by Professor Faraday in his 'Thoughts on Ray Vibrations.' The electromagnetic theory of light, as proposed by him, is the same in substance as that which I have begun to develop in this paper, except that in 1846 there were no data to calculate the velocity of propagation." Maxwell then sets out new equations to express the relations between the electric and magnetic displacements in the medium and the forces to which they give rise. He not only accepts the idea of Faraday that a

moving electric charge (as on a charged body in motion)
acts magnetically as an electric current,—a proposition at
that time unsupported by any experimental demonstration
— but goes further and maintains that there is also a
magnetic action produced during the production or release
of an electric displacement in a dielectric medium : in
fact, that displacement-currents in non-conductors produce,
while they last, exactly the same magnetic action as the
equivalent conduction-current would produce.   He finds
that if magnetic methods of measurement are adopted,
the unit of electricity arrived at has a certain value, while
if purely electrical methods are used the unit has a different
value.   The relation between these two units was found
to depend on the "electric elasticity" of the medium,
and to be a velocity ; namely, that velocity with which an
electromagnetic disturbance is propagated in space.   This
velocity had already been determined as a ratio of units by
Weber and Kohlrausch, who found it to be $3·19 \times 10^{10}$
centims. per second.   The velocity of apparent propagation
of an electric disturbance along a wire had previously been
roughly determined by Wheatstone at a somewhat higher
figure.   Commenting on Weber's result Maxwell proceeds:—
" This velocity is so nearly that of light, that it seems we have
strong reason to conclude that light itself (including radiant
heat, and other radiations, if any) is an electromagnetic
disturbance in the form of waves propagated through the
electromagnetic field according to electromagnetic laws.   If
so, the agreement between the elasticity of the medium as
calculated from the rapid alternations of luminous vibrations, and as found by the slow processes of electrical
experiments, shows how perfect and regular the elastic
properties of the medium must be when not encumbered
with any matter denser than air.   If the same character of
the elasticity is retained in dense transparent bodies, it
appears that the square of the index of refraction is equal to
the product of the specific dielectric capacity and the
specific magnetic capacity.   Conducting media are shown
to absorb such radiations rapidly, and therefore to be
generally opaque."   These two conclusions Maxwell himself

attempted to verify, and pointed out an apparent exception
in the case of electrolytes, which conduct and yet are
transparent.   In Maxwell's theory every electromagnetic
wave must consist of two kinds of displacements both transverse to the direction of propagation, and at right angles to
one another, one being an electrostatic displacement, the
other a magnetic displacement.   In this feature Maxwell's
theory reconciles the conflicting views of Fresnel and
MacCullagh respecting the relation of the displacements to
the "plane of polarisation."   It is now known that the
electric displacements are at right angles to that plane and
agree with the Fresnel vibrations ; whilst the magnetic displacements are in the plane of polarisation as required by
the theory of MacCullagh.   When, in 1874, Maxwell
published his Treatise on Magnetism and Electricity, he
had already attempted a further verification of the theory
by means of a new determination of the ratio of the units.
During the next ten years British physicists were busy
following out the applications of the theory, and testing
its truth in particular instances.   Lord Rayleigh showed
that it led much more readily than the old elastic-solid
theory of light to the equations for double refraction, and to
the explanation of the scattering of light (as in the blue of
the sky) by small particles.   FitzGerald applied it to the
problems of the reflexion and refraction of light.   J. J.
Thomson undertook a new determination of the ratio of
the units.   Ayrton and Perry pursued a similar investigation by a new method.   The same two observers also
verified the relation of the optical and dielectric properties
in the case of gases as required by the theory, and examined the anomalies presented by ice and ebonite.
Poynting and Heaviside independently deduced from
Maxwell's theory the proposition that the energy of an
electric current travels by the medium and not by the
wire.   Hopkinson investigated the relation between the
refractive index of a number of substances and their
dielectric inductivity ; and found some notable deviations
from the values required by theory.   Lodge added a number
of important considerations, and produced mechanical

models in illustration of Maxwell's ideas. The present author investigated the opacity of tourmaline in relation to its conductivity; and found also, in accordance with Maxwell's views, that the conductivity of the double iodide of mercury and copper increases when it is raised to the temperature at which its opacity to light is suddenly augmented. Even more important, because independent of Maxwell's theory, Dr. Kerr in 1876 and 1877 discovered by direct experiment new relations between light and magnetism and between light and electrostatic strain, effects which Faraday had suspected, but sought in vain to discover. Lastly, FitzGerald had, in 1879 and 1883, suggested means of starting electromagnetic waves in the ether.

By the year 1884 all British physicists, except perhaps Lord Kelvin, who had just then been elaborating an independent spring-shell theory of the ether as an improvement on the elastic-solid theory, had accepted Maxwell's theory. Three years later Lord Kelvin gave his adhesion. On the Continent it was, however, barely recognised. In France it was quite ignored until Mascart and Joubert gave some account of it in their treatise on electricity. In Germany it was not quite so entirely neglected. Von Helmholtz appears to have been early drawn to study it, and himself evolved a new theory of dielectric action on similar lines. Later (in 1893) he applied the electro-magnetic theory to explain anomalous refraction and dispersion (see Appendix to Lecture III., p. 100, above). It was von Helmholtz who first drew the attention of Hertz to the possibility of establishing a relation between electro-magnetic forces and dielectric polarisation. Boltzmann had also attempted to verify Maxwell's theory with respect to the relation between the optical and dielectric properties of transparent substances. But, for the rest, Maxwell's theory was practically ignored. Boltzmann himself wrote in 1891 : " The theory of Maxwell is so diametrically opposed to the ideas which have become customary to us that we must first cast behind us all our previous views of the nature and operation of electric forces before we can enter into its portals." Wiedemann appears to have deemed the discrepancies observed by Hopkinson and Boltzmann as sufficient to call in question the validity of the theory, of which little notice is taken in the volumes of the 1885 edition of *Die Lehre von der Elektricität.* (Fleming has in 1897 shown that these discrepancies disappear when the substances are cooled in liquid oxygen to about − 180° C.)

In 1886 Lodge, investigating the theory of lightning conductors, carried out a long series of experiments on the discharge of small condensers, leading him to the observation of electric oscillations and of the travelling of electric waves as guided by wires. Hertz taking up the problem put to him by von Helmholtz, threw himself into investigating the influence of non-conducting media on the propagation of electric sparks. By March 1888 he had succeeded not only in producing electric oscillations and electric waves by the apparatus described above (Fig. 114, p. 215), but in demonstrating that these waves could be reflected and refracted like ordinary light.

The result of the publication of Hertz's work was immediate and widespread. To those continental physicists who had hitherto ignored Maxwell's theory, or who were unaware of the proofs accumulated by British physicists, Hertz's work was nothing short of a revelation. Scientific Europe precipitated itself upon the production of electric oscillations, as if eager to make up lost headway. The revelation was the more significant, since for those who had not accepted the ideas of Faraday and Maxwell as to action in the medium, it meant the abandonment of all the other electrical theories then extant which were based on the now untenable principle of action at a distance. None the less heartily was Hertz's work welcomed in England by those who were already disciples of Maxwell. They saw in it the crowning proofs of a theory which on other grounds they had already accepted as true. To adopt Oliver Lodge's words, by the end of 1888 the science of electricity had definitely annexed to itself the domain of optics, and had become an imperial science.

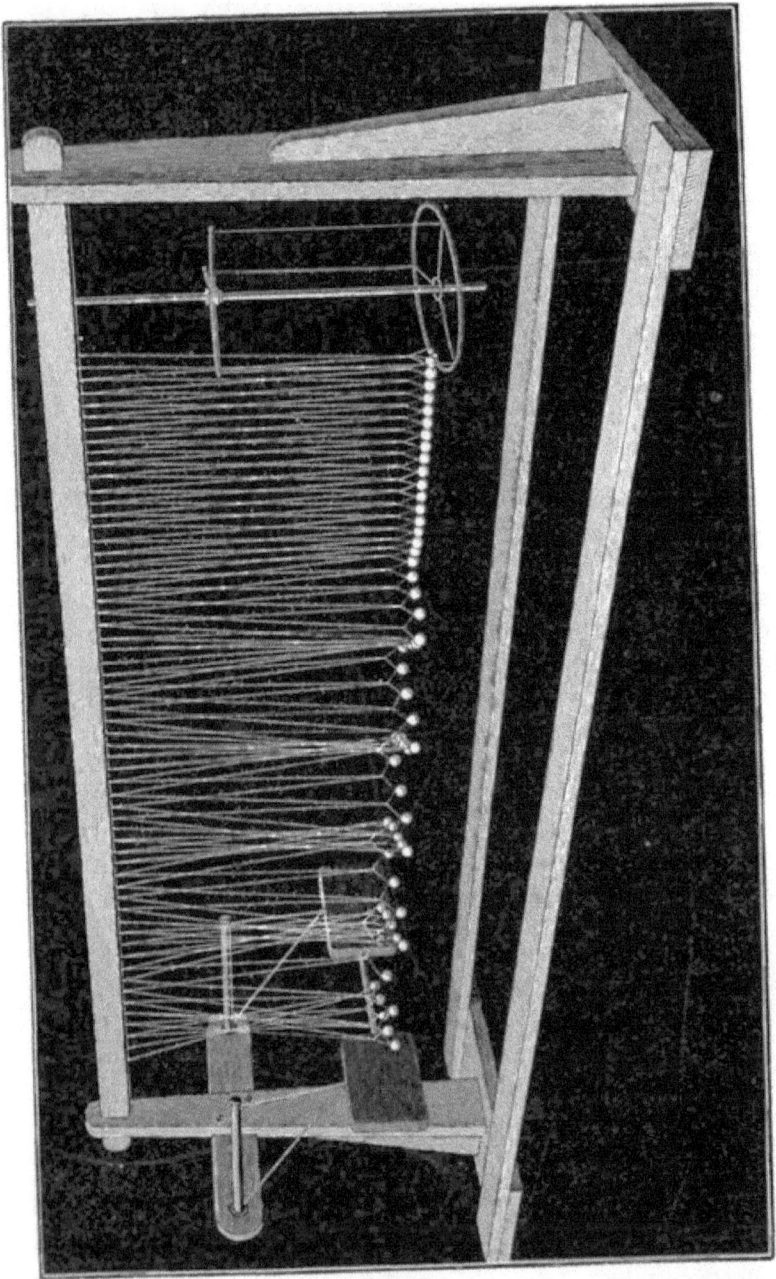

Since 1888 much has been done to complete the experimental verification of the complete analogy between electric waves and waves of light. Of these the more important are briefly noticed on pp. 221-229 above. Suffice it here to say that there is no known physical property possessed by waves of ordinary light that has not been found to be also correspondingly a property of the longer invisible waves produced by purely electric means. By reason of their greater wave-length they will pass through many substances opaque to ordinary light, such as stone and brick walls, and through fogs and mists.

A Hertz-wave Model

Subsequently to the delivery of these lectures, and while this volume was preparing for the press, the author devised a wave-motion model to illustrate mechanically the propagation of a wave from a Hertz oscillator to a Hertz resonator. This apparatus, which is depicted in Fig. 129, should be compared with Fig. 114, p. 215. In this model the "oscillator" is a heavy mass of brass hung by cords, and having a definite time of swing. The "resonator" is a brass circle hung at the other end of the apparatus by a trifilar suspension. They are adjusted by lengthening or shortening the cords so as to have identical periods of oscillation. Between them, to represent the intervening medium and transmit the energy in waves, is a row of inter-connected pendulums (on a plan somewhat similar to one suggested in 1877 by Osborne Reynolds) consisting each of a lead bullet hung by a V thread, the separate Vs overlapping one another so that no bullet can swing without communicating some of its motion to its next neighbour. On drawing the oscillator aside and letting it go it sets up a transverse wave which is propagated along the row of balls in a manner easily followed by the eye, and which, on reaching the resonator at the other end, sets it into vibration.

LECTURE VI

RÖNTGEN LIGHT

So many erroneous accounts have appeared, chiefly in photographic journals, written by persons unacquainted with physical science, that it seems worth while in beginning a lecture on the subject of Röntgen's rays to state precisely how Röntgen's discovery was made, in the language in which he himself has stated it.

" Will you tell me," asked Mr. H. J. W. Dam in an interview [1] with Prof. Röntgen in his laboratory at Würzburg, " the history of the discovery ? "

[1] McClure's *Magazine*, vol. vi., p. 413.

"There is no history," he said. "I had been for a
long time interested in the problem of the kathode rays
from a vacuum tube as studied by Hertz and Lenard.
I had followed theirs and other researches with great
interest, and determined, as soon as I had the time,
to make some researches of my own. This time I
found at the close of last October [1895]. I had been
at work for some days when I discovered something
new."

"What was the date ?"

"The 8th of November."

"And what was the discovery ?"

"I was working with a Crookes's tube covered by a
shield of black cardboard. A piece of barium platino-
cyanide paper lay on the bench there. I had been
passing a current through the tube, and I noticed a
peculiar black line across the paper."

"What of that ?"

"The effect was one which could only be produced,
in ordinary parlance, by the passage of light. No light
could come from the tube because the shield which
covered it was impervious to any light known, even that
of the electric arc."

"And what did you think ?"

"I did not think; I investigated. I assumed that
the effect must have come from the tube, since its
character indicated that it could come from nowhere
else. I tested it. In a few minutes there was no doubt
about it. Rays were coming from the tube, which had
a luminescent effect upon the paper. I tried it successfully at greater and greater distances, even at two metres.

It seemed at first a new kind of light. It was clearly something new, something unrecorded."

"Is it light ?"

"No." [It can neither be reflected nor refracted.]

"Is it electricity ?"

"Not in any known form."

"What is it ?"

"I do not know. Having discovered the existence of a new kind of rays, I of course began to investigate what they would do. It soon appeared from tests that the rays had penetrative power to a degree hitherto unknown. They penetrated paper, wood, and cloth with ease, and the thickness of the substance made no perceptible difference, within reasonable limits. The rays passed through all the metals tested, with a facility varying, roughly speaking [inversely], with the density of the metal. These phenomena I have discussed carefully in my report [1] to the Würzburg Society, and you will find all the technical results therein stated."

Such was Röntgen's own account given by word of mouth. It is entirely borne out by the fuller document, in which in quiet and measured terms Röntgen described to the Würzburg Society his discovery under the title "On a new kind of Rays," and which was the first announcement to the scientific world.

Now you will note that in the whole passage I have read describing the discovery, there is not a word about photography from beginning to end. Photography

[1] Ueber eine neue Art von Strahlen (Vorläufige Mittheilung), von Dr. Wilhelm Konrad Röntgen. (Sitzungsberichte der Würzburger physik-medic. Gesellschaft, 1895.)

played no part in the original observation. No photographic plate or sensitised paper was employed. The
discovery was made by the use of the luminescent screen,
the acquaintance of which you made (if you did not know
of it before) at my fourth lecture, when we were dealing
with ultra-violet light. On that occasion I showed you
a card partly covered with platino-cyanide of barium
which has been in my possession since 1876. When
exposed to invisible ultra-violet light it shone in the
dark. No one who has ever used such a luminescent
screen can blunder into mistaking it for a photographic
plate. Such a screen—a piece of paper covered with the
luminescent stuff [1]—was Röntgen using in his investigations. And as luminescent screens are not things to
be found lying about by accident, it is evident that
its presence on the bench in Röntgen's laboratory on
8th November, 1895, when he was deliberately investigating the phenomena observed by Lenard, was in no
sense accidental. That you may the better understand
the precise nature of Röntgen's discovery, we will
repeat the observation with the appliances now at our
disposal.

Before you stands a Crookes's tube, which I can at
any moment stimulate into activity by passing through
it an electric spark from a suitable induction-coil. It
shines with visible light, the glass glowing with a
beautiful greenish-gold fluorescence. To stop off all

[1] It is interesting to note that Lenard's investigations of 1894
were conducted by the aid of a luminescent screen composed of
paper impregnated with the wax-like chemical called pentadecyl-
paratolylketone.

R

visible light, I place over the Crookes's tube this case
made of black cardboard, which cuts off not only the
visible rays of every sort, but also cuts off the invisible
rays of the infra-red and ultra-violet sorts. On the table,
just below the tube, lies a sheet of paper covered with
platino-cyanide of barium—in fact, a luminescent screen.
And, on passing the electric discharge through the shielded
Crookes's tube you will all see that this luminescent sheet
at once shines in the dark ; while across it—as those
who are near may observe—there falls obliquely a dark
line which is simply a shadow of a small support that
stands between the tube and the screen.   Something
evidently is causing that sheet of luminescent paper to
light up.   Can the effect come from anywhere else than
from the tube ? Try by interposing things, and see
whether they cast shadows on the paper.   The nearest
thing at hand is a wooden bobbin, on which wire is
wound.   If I interpose it, it casts a shadow on the paper.
But looking at the shadow one notices, curiously enough,
that while the wire casts a decided shadow, the wood
casts scarcely any.   I hold up the screen that you may
see the shadow more plainly.   Yes ! there is something
coming from that tube which causes the screen to light
up, and which casts on the screen shadows of things
held between tube and screen.   This light—if light
it be—comes from the tube.   But is it light ? Light, as
we know it, cannot pass through black cardboard.   If it
be light it is light of some wholly new and more penetrative kind.   I move away, still holding the screen in
my hand, to greater distances.   Here, two metres away,
the screen still shines, though less brilliantly.   And,

note, it shines whether its face or its back be presented toward the tube. The rays, having penetrated
the shield of black cardboard that encloses the tube, can
also penetrate the paper screen from the back, and make
the chemically-prepared face shine.   Let us follow
Röntgen farther as he investigated the penetrative power
of the rays.   I interpose a block of wood against which
a pair of scissors has been fixed by nails.   You can see
on the screen the shadow of the scissors; the light
passes through the wood, though not so brightly, for the
wood intercepts some of the rays.   Paper, cardboard,
and cloth are easily penetrated by them.   The metals
generally are more opaque than any organic substance,
and they differ widely amongst one another in their
transparency.   Thin metal foil of all kinds is more or
less transparent; but when one tries thicker pieces they
are of different degrees of opacity.   Ordinary coins are
opaque.   A golden sovereign, a silver shilling, and a
copper farthing are all opaque, but the lighter metals
such as tin, magnesium, and aluminium, notably the latter,
are fairly transparent.   Here is my purse of leather with
a metal frame.   I have but to hold it between the tube
and the screen to see its contents—two coins and a
ring—for leather is transparent to these rays.   A sheet
of aluminium about the twentieth of an inch thick,
though opaque to every other previously-known kind of
light is for this kind of light practically transparent.   On
the other hand lead is very opaque.   Röntgen found
opacity to go approximately in proportion to density.
It is now found that those metals which are of the
greatest atomic weight are the most opaque to Röntgen

light. Uranium, the atomic weight of which is 240, is the most opaque; whilst lithium, whose atomic weight is only 7, and which will readily float on water, is exceedingly transparent. In fact I have never yet got a good shadow from lithium. This relation extends not only to the metals themselves but to their compounds. Thus the chloride of lithium is more transparent than the chloride of zinc or than the chloride of silver. Finding that the denser constituents were the more opaque, and that while glass and stone are tolerably opaque such substances as gelatine and leather were comparatively transparent, it occurred to Röntgen that bone would be more opaque than flesh—and so it proved: for interposing the hand between the tube and the screen we find that while the flesh casts a faint shadow the bones cast a much darker one, and so we are able to see upon the luminescent screen, in the darkness, the shadow of the bones of the hand, and of the arm. This is truly seeing the invisible.

But now the investigation took another turn. So far there has been no mention of photography. But the peculiar penetrative light having been discovered, and the shadows having been seen on the luminescent screen, it was a pretty obvious step to register these shadows photographically. For, as was already well known in the case of ultra-violet light, the rays that stimulate fluorescence and phosphorescence are just those rays which are most active chemically and photographically. Hence it was to be expected that these new rays would affect a photographic plate. This Röntgen proceeded to verify. He obtained a photograph of a set of metal

weights that were shut up in a wooden box.   Also of a
compass, showing the needle and dial through the thin
brass cover.   He then put his tube under a wooden-
topped table ; and laying his hand on the table above
it, and poising over it a photographic dry-plate, face
downwards, he threw upon the plate, by light which
passed upwards through the table top, a shadow of his
hand.   So for the first time he succeeded in photographing the bones of a living hand.   It was the photography
of the invisible.   But, note, even here there is no " new
photography."   The only photography in the matter is
the well-known old photography of the dry-plate, which
must first be exposed and afterwards developed in the
dark-room.

And now, though it anticipates somewhat the course
of this lecture, since the process of photographic development in the dark-room requires a little time, I will proceed to take a few photographs which will then be taken
to the dark-room to be developed, and will afterwards
be brought back and shown you upon the screen by
means of the lantern.

[In the experiments which followed photographs were
taken of the hands of a boy and of a girl, also shadows
cast by sundry gems, including a fine Burmese ruby, a
sham ruby, a Cape diamond, and an Indian diamond.]

Retracing our steps in the order of discovery I must
at once take you back, nearly two hundred years, to
the time of Francis Hauksbee, when, with the newly
invented electric machine, and the newly perfected air-
pump, the first experiments were made on the peculiar
light produced by passing an electric spark into a partial

vacuum. About that time Europe was nearly as much excited—considering the state of knowledge and the slow means of communication—over the "mercurial phosphorus," as it was last year over the Röntgen rays. This "mercurial phosphorus" was simply a little glass tube, such as that (Fig. 130) which I hold in my hand. It contains a few drops of quicksilver; and the air that otherwise would fill the tube has been mostly pumped out by an air-pump, leaving a partial vacuum. I have but to shake the tube and it flashes brightly with a greenish light. The friction of the mercury against the

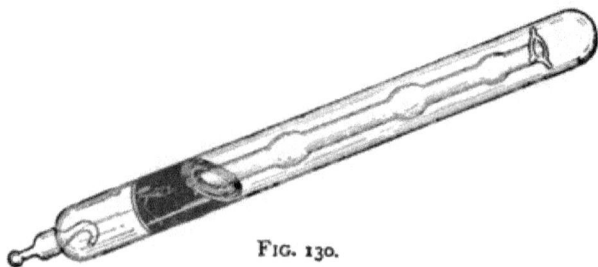

FIG. 130.

glass walls sets up electric discharges, which flash through the residual air, illuminating it at every motion.

While I have been talking to you an air-pump in the basement, driven by a gas-engine, has been at work exhausting a large oval-shaped glass tube. Only perhaps $\frac{1}{300}$ part of the air originally in it remains. On sending through it from top to bottom the sparks from an induction coil, it is filled with a lovely pale crimson glow, which changes at the lower end to a violet-coloured tint. On reversing the connections so as to send the discharge upwards the violet-coloured part is seen at the top. It shows you, in fact, the end

at which the electric discharge is leaving the tube. The
pale glow of this primitive vacuum tube is rich in light
of the ultra-violet kind, which, as you know, readily
excites fluorescence. I have but to hold near it my
platino-cyanide screen for you to observe the rich green
fluorescence. My hand will cast a shadow on the screen
if I interpose it, but there are no bones to be seen in
the shadow. For here there is none of the penetrative
Röntgen light : the fluorescence is due to ordinary ultra-
violet waves, to which flesh and cardboard are quite
opaque. If the tap is turned on to readmit the air you
see how the rosy glow contracts first into a narrowing
band, then into a mere line, which finally changes into
a flickering forked spark of miniature lightning ; and all
is over until and unless we pump out the air again.
Another beautiful effect is shown by use of an exhausted
glass jar, within which is placed a cup of uranium glass,
as described fifty years ago by Gassiot. The discharge
overflows the cup in lovely streams of violet colour,
while the cup itself glows with vivid green fluorescence.
Some thirty years ago vacuum tubes became an article
of commerce, and were made in many complex and
beautiful shapes by the skill of Dr. Geissler of Bonn,
who devised a form of mercurial air-pump [1] for the purpose of extracting the air more perfectly ; though the
degree of vacuum, which sufficed to display the most
brilliant colours when stimulated by an electric discharge,
is far short of that which is requisite in the modern

[1] See the author's monograph on *The Development of the Mercurial Air-Pump*, published in 1888, by Messrs. E. and F. N.
Spon.

vacuum tubes of which I must presently speak. Here
is a Geissler's tube showing wondrous effects when the
spark discharge is passed into it. Strange flickering
striations palpitate along the windings of the glass tubes
which themselves glow with characteristic fluorescence.
Soda-glass fluoresces with the golden-green tint, lead
glass with a fine blue, and uranium glass with a brilliant
green. The violet glow which appears in the bulb at
one end of the tube surrounds the metal terminal by
which the current leaves the tube, and is itself due to
nitrogen in the residual air. Each kind of gas gives its
own characteristic tint. And with any kind of gas within

FIG. 131.

the tube the luminous phenomena are different at different
degrees of exhaustion.

I have here a set of eight tubes, all of the same simple
shape (Fig. 131), but they differ in respect of the degree
of vacuum within them. Platinum wires have been sealed
through the ends of each, one wire *a* to serve as the *anode*
or place where the electric current enters, another wire *k*
to serve as *kathode* or place where the current makes its
exit from the tube. Both anode and kathode are tipped
with aluminium, as this metal does not volatilise so
readily under the electric discharge. The small side-
tube *s* by which the tube was attached to the pump
during exhaustion is hermetically sealed to prevent air
from re-entering. The first tube of the set is full of air
at ordinary pressure, and does not light up at all. The

length between anode and kathode (about 12 inches) is so great that no spark will jump between them. In the second tube the air has been so far pumped away that only about $\frac{1}{5}$ of the original air remains. Across this imperfect vacuum forked brush-like bluish sparks dart. The third tube has been exhausted to about $\frac{1}{20}$ part; that is to say, $\frac{19}{20}$ of the air have been removed. It shows, instead of the darting sparks, a single thin red line, which is flexible like a luminous thread. In the fourth tube the residual air is reduced to $\frac{1}{40}$ or $\frac{1}{50}$ part; and you note that the red line has widened out into a luminous band from pole to pole, while a violet mantle makes its appearance at each end, though brighter at the kathode. In the fifth tube, where the exhaustion has been carried to about $\frac{1}{500}$, the luminous column, which fills the tube from side to side, has broken up into a number of transverse striations which flicker and dance; the violet mantle around the kathode has grown larger and more distinct. It has separated itself by a dark space from the flickering red column, and is itself separated from the metal kathode by a narrow dark space. The degree of exhaustion has been carried in the sixth tube to about $\frac{1}{10000}$: and now the flickering striations have changed both shape and colour. They are fewer, and whiter. The light at the anode has dwindled to a mere star; whilst the violet glow around the kathode has expanded, and now fills the whole of that end of the tube. The dark space between it and the metal kathode has grown wider, and now the kathode itself exhibits an inner mantle of a foxy colour, making it seem to be dull and hot. The glass, also, of the tube shows a tendency to emit a green fluorescent light at the kathode end. In the seventh tube the exhaustion has been pushed still farther, only about $\frac{1}{50000}$ part of the original air being left. The luminous column has subsided into a few greyish-white nebulous patches. The dark space around the kathode has much expanded, and the glass of the tube exhibits a yellow-green fluorescence. In the eighth tube only one or two millionths of the original air are present; and it is now found much more difficult to pass a spark through the tube. All the internal flickering clouds and striations in the residual gas have disappeared. The tube looks as if it were quite empty: but the glass walls shine brightly with the fine golden-green fluorescence, particularly all around the kathode. If we had pushed the exhaustion still farther, the internal resistance would have increased so much that the spark from the induction coil would have been unable to penetrate across the space from anode to kathode.

To attain such high degrees of exhaustion as those of the latter few tubes recourse must be had to mercurial air-pumps; no mechanical pump being adequate to produce sufficiently perfect vacua. The Sprengel pump, invented in 1865 by Dr. Hermann Sprengel, is an admirable instrument for the purpose. But it was modified and greatly improved[1] about 1874 by Mr. Crookes,

[1] These improvements comprised the following :—A method of lowering the supply-vessel to refill it with the mercury that had run through the pump ; the use of taps made wholly of platinum to ensure tightness ; the use of a spark-gauge to test the perfection of the vacuum by observing the nature of an electric spark in it ; the use of an air-trap in the tube leading up to the pump-head ; the

whose form of pump is shown in Fig. 132.   Mercury is
placed in a supply-
vessel, which can be
raised to drive the
mercury through the
pump, and lowered,
when empty, to be
refilled. This vessel
is connected by a
flexible indiarubber
tube to the pump,
which consists of
glass-tubes fused to-
gether.   From the
pump-head the mer-
cury falls in drops
down a narrower
tube, called the fall-
tube, and each drop
as it falls acts as a
little piston to push
the air in front of it,
and so gradually to
empty the space in
the farther part of the tube.   A drying-tube, filled with

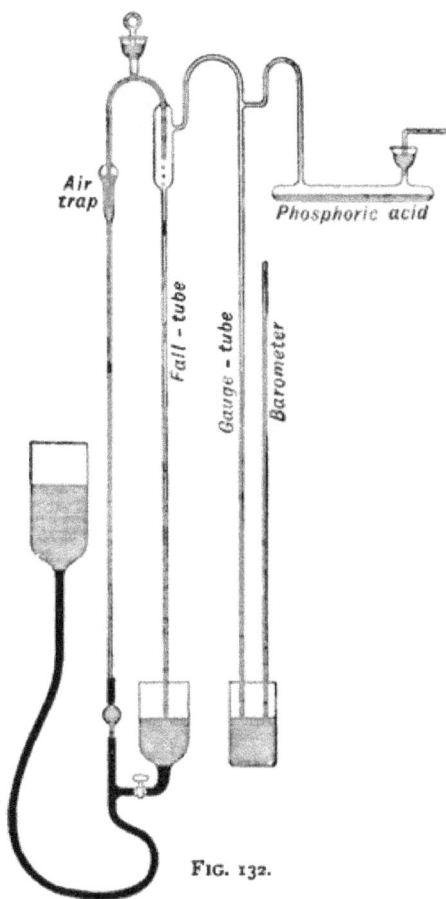

Fig. 132.

method of connecting the pump with the object to be exhausted,
by means of a thin, flexible, spiral glass tube ; the method of
cleansing the fall-tube by letting in a little strong sulphuric acid
through a stoppered valve in the head of the pump.   In carrying
out these developments Mr. Crookes was assisted by the late Mr.
C. Gimingham, whose later contributions to the subject are described in the author's monograph on the Mercurial Air-pump.

phosphoric acid to absorb moisture, is interposed between
the pump and the vacuum-tube that is to be exhausted.
It is usual to add a barometric gauge to show the degree
of vacuum that has been reached.

Before you, fixed against the wall, is a mercury-pump
substantially like Fig. 132, but having three fall-tubes
instead of one, so as to work more rapidly. Through
these fall-tubes mercury is dropping freely; the pump
being at the present moment employed in the exhaustion
of a Crookes's tube, which has been sealed to it by a
narrow glass tube. When the exhaustion has been
carried far enough, this narrow pipe will be melted with
a blow-pipe, so as to seal up the tube and enable it to
be removed from the pump.

It was with such a pump as this that Crookes was
working from 1874 to 1875 in the memorable researches
on the repulsion caused by radiation, which culminated
in the invention of that exceedingly beautiful apparatus
the *radiometer*, or light-mill, which we were using in my
last lecture. From that series of researches Mr. Crookes
was led on to another upon the phenomena of electric
discharge in high vacua. Professor Hittorf of Münster
had already done some excellent work in this direction.
He had noted the golden-green fluorescence around the
kathode when the exhaustion was pushed to a high
degree; and he had found that this golden glow, unlike
the luminous column which at a lower exhaustion fills
the vacuous tube, refuses to go round a corner. He
had even found that it could cast shadows, owing to its
propagation in straight lines.

Starting at this point on his famous research, Crookes

investigated the properties of this kathode light, and
found it to differ entirely from any known kind of radiation.   It appeared to start off from the surface of the
kathode and to move in straight lines, penetrating to
a definite distance, the limit of which was marked by
the termination of the " dark space," according to the
degree of exhaustion, and causing the bright fluorescence
when the exhaustion was carried so far that the dark
space expanded to touch the walls.   Acting on this hint
he proceeded to construct tubes in which the kathode,
instead of being as previously a simple wire, was

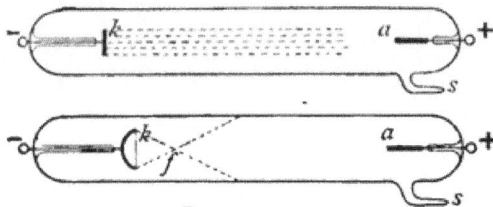

FIGS. 134, 135.

shaped as a flat disk, or as a cup (Figs. 134, 135).   From
the flat disk the kathode rays streamed backwards in a
parallel beam.   Crookes regarded these kathode streams
as flights of negatively-electrified molecules shot backwards from the metal surface.   Doubtless such flying
molecules of residual gas there are ; and they take part
in the phenomenon of discharge, bombarding against the
opposite wall of the tube.   There are, however, strong
reasons for thinking that the kathode rays are not
merely flights of " radiant matter," but that the flying
molecules are accompanied by ether-waves or ether-
motions which cause the fluorescence on the walls of the
tube.   Be that as it may, Crookes found the kathode

rays to be possessed of several remarkable properties. Not only could they excite fluorescence and phosphorescence to a degree previously unknown, but they exercised a mechanical force against the surfaces on which they impinged. They cast shadows of objects interposed in their path; and were capable of being drawn aside by the influence of a magnet, just as if they were electric currents.

Here are some Crookes's tubes which display the luminescent effects. At the top of the first is a small flat disk of aluminium to serve as kathode. From it shoots downward a kathode-beam upon a few Burmese rubies fixed below. They glow with a crimson tint more intense than if they had been red-hot. In a similar tube is a beautiful phenakite,[1] looking like a large diamond. When exposed to the kathode rays it luminesces with a lovely pale blue tint. In the third is placed a common whelk shell, which has been lightly calcined. As the kathode rays stream down upon it it lights up brilliantly. And, after the electric discharge has been switched off, the shell continues for some minutes to phosphoresce with a persistent glow.

In the next tube, which contains a sheet of mica painted with a coat of sulphate of lime so that you may better see the bright trace of its luminescence, a narrow kathode ray is admitted through a slit at the bottom, and extends in a fine bright line upwards. Holding a

---

[1] A species of white emerald found in the Siberian emerald mines, and often sold in Russia as a Siberian diamond. It is not so brilliant as a diamond, though much more rarely met with.

magnet near it, I draw the kathode ray on one side,
illustrating its deflectibility.

To illustrate the mechanical effect of the kathode
rays I take a Crookes's tube, having at its ends flat
disks of metal as electrodes. Between them is a nicely-
balanced paddle-wheel, the axle of which runs upon a
sort of little railway. On sending the spark from the
induction-coil through the tube the little wheel is driven
round and runs along the rails. Its paddles are driven
as if a blast issued from the disk which serves as kathode.
On reversing the current its motion is reversed.

Here (Fig. 136) is a Crookes's tube of a pear shape,
having a piece of sheet-
metal in the form of a
Maltese cross set in the
path of the kathode
rays. See what a fine
shadow the cross casts
against the broad end
of the tube; for the
whole end of the tube

FIG. 136.

glows with the characteristic golden-green luminescence,
except where it is shielded from the rays by the metal
cross.

With this tube I am able to show you a most interesting and novel experiment discovered only a few days
ago by Professor Fleming. If you surround the tube
with a magnetising coil through which an electric current
is passed, the magnetic field produces a remarkable
effect on the shadow. Instead of pulling it on one side
(as a horse-shoe magnet would do), the magnetising

coil causes the cross to rotate on itself, and at the same
time to grow smaller. To show the effect more conveniently I have put the magnetising coil not around
the tube itself, but around an iron core beyond the end
of the tube. So I am able to diminish or augment the
effect by simply moving the tube away from the iron
core, or by bringing it nearer. As I move it up, the
shadow of the cross contracts, and grows smaller but
brighter. It also twists round and turns completely over
top for bottom as it vanishes into a mere point. But
just as it vanishes you see its place taken by a second
large shadow, which, as I push the tube still closer to
the magnetised core, grows brighter and also turns
round and contracts as its predecessor did. Its arms
are more curved than those of the first cross. At the
same moment when the second shadow-cross appears a
third shadow makes its appearance as a distorted
annular form against the walls of the tube between the
metal cross and the kathode. Its position is such that
the shadow seems to have been cast as by rays diverging
from the other end of the tube. As yet we know not
the explanation of these remarkable facts.

The last tube of this set that illustrates Crookes's
researches has as kathode a large hollow cup of
aluminium at the bottom (Fig. 137). This concave
kathode focuses the kathode rays by converging them
to a point in space a little above the centre of the tube.
Crookes found that if the kathode rays were in this way
focused upon anything, they produced great heat.
Glass was melted, diamonds charred, platinum foil
heated red-hot and even fused by the impact of the

concentrated kathode stream.   In the focusing-tube
now before you—an old one, made more than ten years
ago—there is a piece of thin platinum foil hung in the
tube to be heated by the rays.   But it has become displaced and no longer hangs in the focus.   Yet by holding a small horse-shoe magnet outside the tube to
deflect the rays a little, I can displace the focus until it

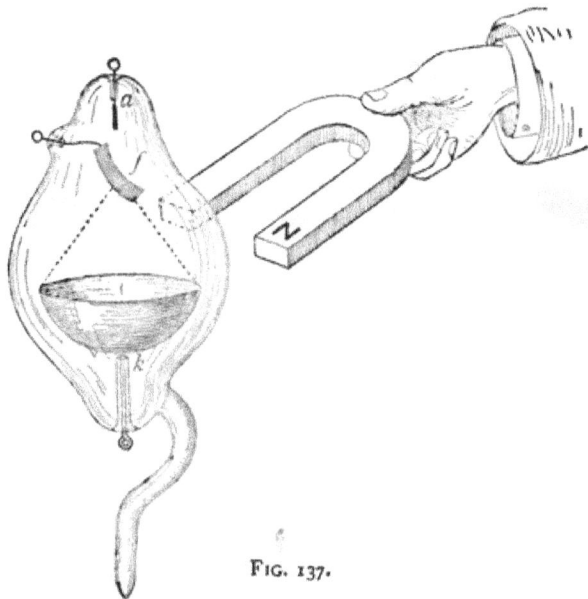

Fig. 137.

falls upon the surface of the platinum foil, which you
now see is raised to a bright red heat.

Since the date, now nearly twenty years ago, when
these most beautiful and astonishing observations were
made by Crookes, there has been much speculation as
to the nature of these interior kathode rays ; their properties were so extraordinarily different from anything
in the nature of ordinary light that even the name " ray "

S

as applied to them seemed out of place. Crookes's own term, "radiant matter," was objected to as necessarily implying their material nature; and yet no other explanation of them seemed reasonable than Crookes's own suggestion that they consisted of flights of electrified particles. It was supposed that they could only exist in a vacuum tube under an exceedingly high condition of exhaustion.

However in 1894 Dr. Philipp Lenard, acting on a hint afforded by an observation of Professor Hertz[1]

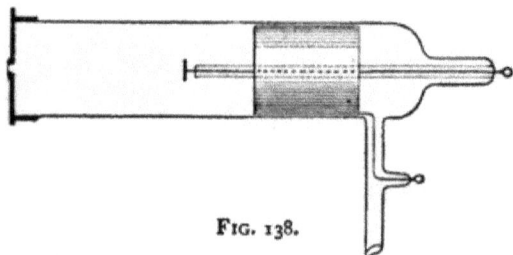

FIG. 138.

succeeded in bringing out the kathode rays into the air at ordinary pressure. For this purpose he fitted up a tube with a small window of thin aluminium foil opposite the kathode, as shown in Fig. 138. The general form of tube was the same as that previously used by Hertz, namely, cylindrical, with a small kathode disk on the end of a central wire, protected by an inner glass tube. The anode was a cylindrical metal tube surrounding the kathode. Upon the further end of the

[1] Hertz noticed that when a very thin metal film was interposed inside the Crookes tube, the glass still fluoresced under the kathode discharge. He found this still to be the case when the film was replaced by a piece of thin aluminium foil which was quite opaque to light.

tube was cemented a brass cap, having at its middle a small hole covered with aluminium foil $\frac{1}{10000}$ inch thick. Through this "window," when the tube was highly exhausted, there came out into the open air rays which, if not actual prolongations of the kathode rays, are closely identified with them. They can be deflected by a magnet—though in varying degrees depending on the internal vacuum. They can excite luminescence. Lenard explored them by using a small luminescent screen of paper covered with a chemical called penta-decylparatolylketone. He found them to be capable of affecting a photographic dry-plate; and studied both by the luminescent screen and by the photographic plate their power of penetrating materials. He found that air at ordinary pressure was not very transparent, acting toward them as a turbid medium. He found them to pass through thin sheets of aluminium and even of copper. He also caused them to affect a photographic plate that was completely enclosed in an aluminium case, and to discharge an electroscope enclosed in a metal box. All this work was done in 1894 and 1895 and duly published. Though it excited no public notice, it was regarded by physicists as of very great importance.

As you were told at the beginning in Röntgen's own account of the matter, his research began with the deliberate aim of reinvestigating the problem of the emission of kathode rays from the vacuum tube as studied by Hertz and Lenard. So as Lenard had done, he employed a luminescent screen to explore the rays, and used a Crookes tube (Fig. 139) of a form closely resembling Lenard's, and indeed identical with that

previously employed by Hertz. The end opposite the
kathode was simply of glass, without any brass cap or
aluminium window. Thus prepared he found what I
have already described, those mysterious rays which
with characteristic modesty he described as " X-rays,"
but which will always be best known as Röntgen's rays.
They are not kathode rays, though caused by them.
Kathode rays will not pass through glass, and are deflected by a magnet. Röntgen rays will pass through
glass and are not deflected by a magnet. They seem

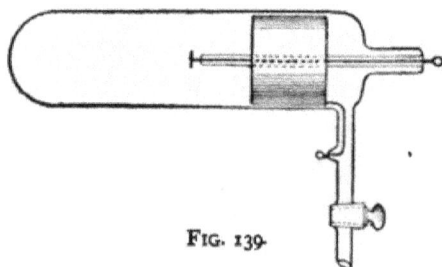

FIG. 139.

indeed to be formed by the destruction of the kathode
rays, having for their origin the spot where the kathode
rays strike against any solid object, best of all against
some heavy metal such as platinum or uranium.
Neither are they ordinary light of either the infra-red
or of the ultra-violet kind, though they resemble the
latter in their chemical activity and in so freely exciting
luminescence. But ultra-violet light can, as we have seen
in previous lectures, be reflected, refracted, and polarised,
while Röntgen light cannot.[1] Nor are Röntgen's rays

[1] Reflexion there is, but not of a regular kind ; the supposed
cases of true reflexion announced by Lord Blythswood and others
belong to the category of myths. There is diffuse reflexion of

the same thing as Lenard's rays; for the latter are in various degrees deflectible by the magnet; and air is toward them relatively much more opaque than it is for Röntgen's rays. Röntgen seems to have been fortunate in having the means of producing the most perfect exhaustion by his vacuum pump: for on the perfection of the vacuum more than on any other detail does the successful production of the Röntgen rays depend. The vacuum, which is abundantly good enough to evoke luminescence, or to show the shadow of the cross, or to produce the heating at the focus, or to drive the " molecule mill," does not suffice to generate the Röntgen rays. For this last purpose the exhaustion must be carried to a higher point—to a point so high indeed that the tube is on the verge of becoming non-conductive.

Röntgen rays from polished metals, particularly from zinc, just as there is diffuse reflexion of ordinary light from white paper. As to refraction, Perrin in Paris, and Winkelmann in Jena, have independently found what they think evidence of feeble refraction through aluminium prisms. But the deviation (which is towards the refracting edge) is so excessively small as to be scarcely distinguishable from mere instrumental errors. Polarisation has been looked for by many skilled observers, using many materials including tourmalines. Only one success has been alleged, by MM. Galitzine and Karnojitzky, using tourmaline ; but their result has not been confirmed and is probably erroneous. Neither has interference of Röntgen light yet been shown to be possible. Several observers have professed that they have obtained diffraction fringes from which the wave-length of the Röntgen rays could be measured. But some of these measurements show a wave-length greater than that of red light, and others less than that of ordinary ultra-violet : they are probably all due to some unnoticed source of error. None of them can be accepted without subsequent confirmation by other experimenters, and this is not yet forthcoming.

Here let me say a word about the man himself and his material surroundings. Still in the prime of life, at the age of fifty-one, Professor Wilhelm Konrad Röntgen had already made himself a name among physicists by his work in optics and electricity before the date of the brilliant discovery that gave him wider fame. He occupies the chair of Physics in the University of Würzburg in Bavaria, and lives and works in the physical laboratory of the University. The little town of Würzburg, of 61,000 inhabitants, boasts a university frequented by 1,490 students, and supported with an income of £41,000 a year, of which more than half is contributed by the State. There are 53 professors and 40 assistants. Its buildings comprise a group of laboratories and institutes devoted to chemistry, physiology, pathology, mineralogy, and the like. Its physical laboratory, a neat detached block of buildings, wherein also the professor has his residence, is of modern design. Its equipment for the purpose of research is infinitely better than that of the University of London;[1] and it is

[1] From a Report recently presented to the Convocation of the University of London, it appears that the physical and chemical laboratories of the University are practically non-existent. "There are three rooms at Burlington House which are occasionally used as laboratories during examinations, and for examinational purposes only. The largest of these is a large hall lit from the top. When used as a chemical laboratory, it is fitted up with working benches down the middle and along the two sides, the benches being divided into separate stalls to isolate candidates in their work. It was stated that the middle stalls and benches are taken down when the hall is used for written examinations, and are re-erected when a chemical examination is to be held. In a second hall, also lighted from above, where frequent written examinations are held, temporary arrangements are made whenever an examination in practical

Fig. 140.— Professor W. K. Röntgen.

expected of the professor that he shall contribute to the advancement of science by original investigations. With such material and intellectual encouragements to research as surround the university professor in even the smallest of universities in Germany, what wonder that advances are made in science ? Would that a like stimulus were existent in England. The Professor of Physics in the University of London has made no discovery like that of Professor Röntgen, for the very good reason that the University of London has neither appointed any Professor of Physics, nor built any physical laboratory where he might work. Neither the State nor the municipality has provided it with the necessary funds. Its charter

physics is to be held. A curtain of black cloth slung across one end of the room gave partial obscurity over the tables where photometric and spectroscopic apparatus was placed. The third room, sometimes called the galvanometer room, is a smaller room in the basement, artificially lighted, and used chiefly for printing, except at the times of examinations in practical physics." Such is the melancholy state of things in a University where everything is sacrificed on the altar of competitive examinations.

Bavaria has a population of 6,000,000. It supports the three Universities of Munich, Erlangen, and Würzburg, with a total of over 6,000 students, at a cost of £150,000 a year, of which £93,000 is provided by the State. London, with a population of 5,000,000, has the University of London, a mere Examining Board, to which come up for intermediate and degree examinations about 2,000 students yearly, of whom a large proportion are from the provinces. It has no professors. Its laboratories are in the deplorable position above mentioned. So far from being endowed by the State, it pays in to the State about £16,500 a year, and nominally receives back about £16,280 as a parliamentary grant. It receives no subvention from the municipality. Its library is closed for a large portion of the year, the room being used for examination purposes almost every day.

precludes it from doing anything for science except hold examinations! Perhaps some day London may have a university worthy of being mentioned beside that of Würzburg, which is eleventh only in size amongst the universities of Germany.

Röntgen had so thoroughly explored the properties of the new rays by the time when his discovery was announced, that there remained little for others to do beyond elaborating his work. One point deserves notice; namely, the improvement of the tubes. Röntgen held the view that his rays originated at the fluorescent spot where the kathode rays struck the glass. This led some persons to the idea that fluorescence was advantageous. Several workers, however, discovered about the same time that if the kathode rays were focused upon a piece of metal the emission of Röntgen light became more copious. When studying early last year the conditions under which the rays were produced, I found that the best radiators are substances which do not fluoresce—namely, metals. I found zinc, magnesium, aluminium, copper, and iron to answer; but platinum was better than these, and uranium best of all. Directing the kathode discharge against a target or "antikathode" of platinum fixed in the middle of the tube, I carefully watched, by aid of a luminescent screen, the emissive activity of the surface during the process of exhaustion. After the stage of exhaustion has been reached at which Crookes's shadows are produced, one must go on further exhausting before any trace of Röntgen rays appear. The first luminosity seems to come (as in Fig. 141) from both front and back of the target at once; an oblique line, corresponding to the plane of the " antikathode " or target of metal, being seen on the screen between two partially luminescent regions.   On continuing the exhaustion, the light behind dies out while that in front increases, as in Fig. 142, the rays being emitted copiously right up to the plane of the antikathode.   This

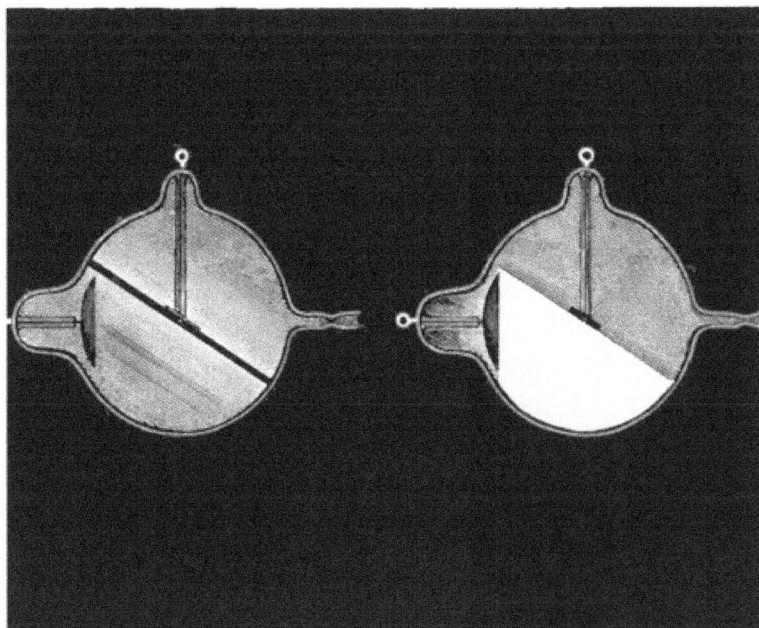

FIG. 141.   FIG. 142.

lateral emission is quite unlike anything in the emission or reflexion of ordinary light, and has to be accounted for in any theory of the Röntgen rays.   I have myself observed[1] that within the tube there are some other rays given off in a similar way, along with the Röntgen rays, but which are not Röntgen rays, for they can be

[1] See *Electrician.*

deflected by a magnet, and more nearly resemble the
kathode rays.   It is these that produce on the glass
wall of the tube a well-marked fluorescence delimited
(as in Fig. 142) by an oblique plane corresponding to
the delimitation of Röntgen rays seen in the fluorescent
screen.   The tube which I used at the beginning of
this lecture, and which we will use again at the close
of the lecture to show you your own bones, is of the
focus type (Fig. 143).   It is of the pattern devised by

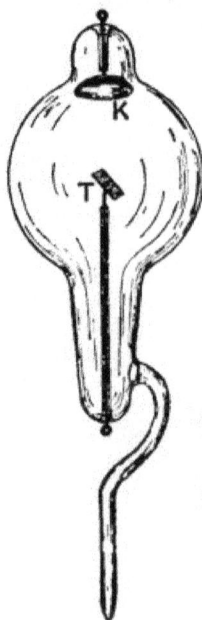

FIG. 143.

Mr. Herbert Jackson, of King's College.   The concentration of the kathode rays upon the little target of
platinum (which often becomes red-hot)
has the advantage not only of allowing
a more copious emission of Röntgen
rays than would be possible if the anti-
kathodal surface were the glass wall,
but also of causing the Röntgen rays
to issue from a small and definite
source so that the shadows cast by
objects are more sharply defined.
Here are two still more recent tubes
(Figs. 144, 145) constructed for me
by Mr. Böhm, in which the focus principle is preserved; but in which there
is the improvement that the anti-
kathode T is not used also as an anode.   It is an
insulated target of platinum, while the anodes are
aluminium rings through which the cone of kathode
rays passes.   These tubes are not liable to blacken, as
is the case with tubes in which the antikathode is also

used as anode. The tube (Fig. 145) has two concave electrodes, either or both of which may be used as

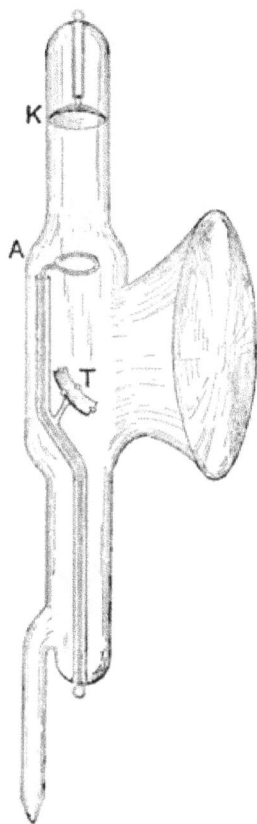

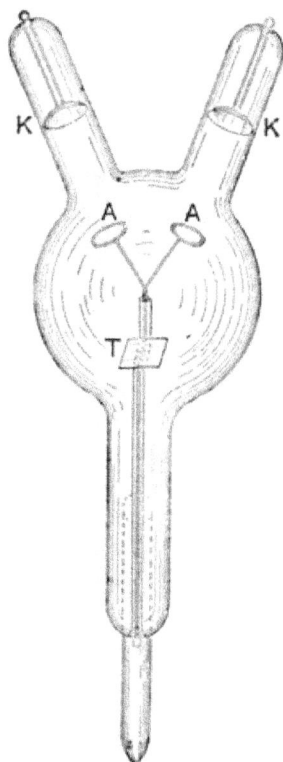

FIG. 144.   FIG. 145.

kathode; it is a convenient form for those cases in which an alternating current is employed.

In another direction many efforts have been made at improvement of the luminescent screen. At first good barium platino-cyanide was not to be procured, and hydrated potassium platino-cyanide was found far superior. But the good barium salt now procurable is quite as luminescent, and is less troublesome to manage. One result of the ignorance which at first prevailed as to the real origin of Röntgen's discovery was that various experimenters up and down the world supposed themselves to have invented something when they took to using fluorescent screens. One man puts a fluorescent screen at the bottom of a pasteboard tube, with a peep-hole lens at the top, and calls it a " cryptoscope." Another, in another part of the globe, puts a fluorescent screen at the bottom of a nice cardboard box furnished with a handle and a flexible aperture to fit to the eyes, and styles it a " fluoroscope." Both are useful; but the only invention in the whole thing is Röntgen's.

Within a few days of the publication of Röntgen's discovery another effect, however, which had escaped Röntgen's scrutiny, was observed by several independent observers. It had been known for several years that when ultra-violet light falls upon an electrically-charged surface it will cause a diselectrification, but only if the surface is negatively charged. Ultra-violet light will not diselectrify a positive charge.[1] But Röntgen rays are found to produce a diselectrification of a metal surface (in air) whether the charge be positive or negative. Here is a convenient arrangement for exhibiting the experiment. An electroscope made on Exner's plan with three leaves—the central one a stiff plate of metal—is charged, and then exposed to Rönt-

---

[1] Ultra-violet light will not diselectrify a *metal* surface in air unless that surface is negatively charged. I have observed a case, however, in which a positively-electrified body was discharged by ultra-violet light, but it was not a metal surface, nor in air.

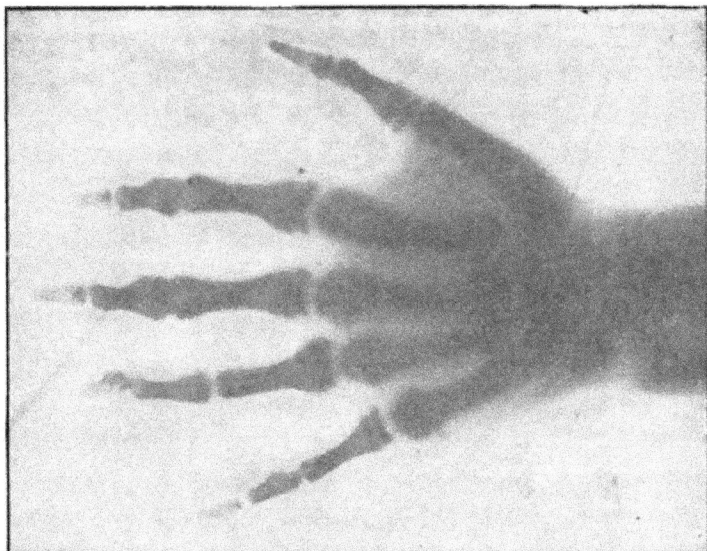

FIG. 148.

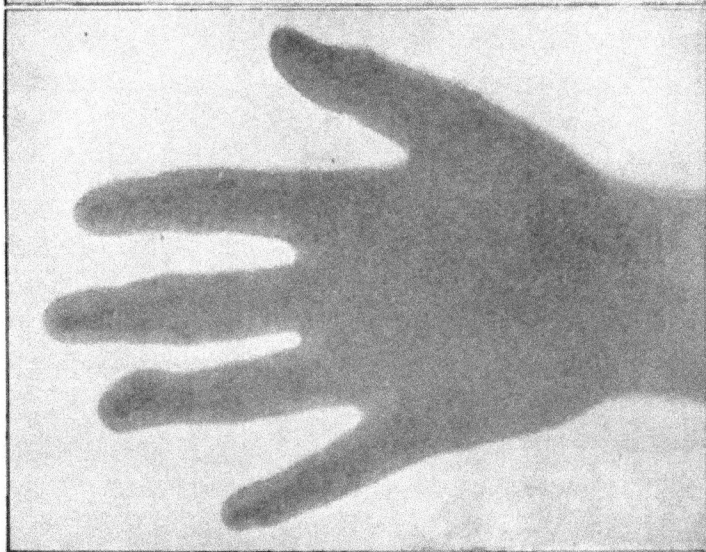

FIG. 147.

gen light.   The three leaves are made of aluminium,
aluminium foil being better than leaf-gold for electro-
scopes.   They are supported within a thin
flask of Bohemian glass entirely enclosed,
except at the top, in a mantle of transparent metallic gauze.   After the leaves
have been charged—either positively by a
rod of rubbed glass, or negatively by a rod
of rubbed celluloid—a metal cap is placed
over the top (Fig. 146).

The leaves, being thus completely surrounded by metal, are effectually screened
from all external electrical influences.   My

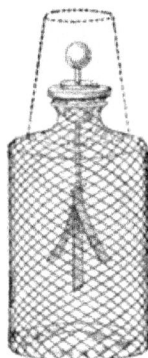

FIG. 146.

electroscope is now charged.   To enable you to see
the effect better, a beam of light is directed upon it,
throwing a magnified shadow of the leaves upon the
white screen.   Then, exposing the electroscope to
Röntgen light from a focus tube situated some 18
inches away, you see the leaves at once closing together,
proving the diselectrification.   It succeeds whether the
charge be positive or negative in sign.

It now only remains for me to exhibit to you the
photographs which were taken at the beginning of this
lecture, and a number of others prepared as lantern-
slides.   In Figs. 147, 148 we have the hand of a poor
child aged thirteen, a patient in St. Bartholomew's
Hospital.   She was brought to my laboratory that the
deformities of her hands might be examined.   The first
of the two plates was insufficiently exposed, with the
result that the bones scarcely show through the flesh at
all.   The second plate was over-exposed, and the rays

T

have penetrated the flesh so thoroughly that only the bones appear.

Fig. 149 is the hand of a child of eleven years old. In a child's hand the bones are not yet completely ossified, their ends being still gelatinous and transparent, so that there seem to be gaps between them. Compare this with the hand of a full-grown man, and you will see how age changes the aspect of the bones.

Fig. 150 is the hand of a full-grown woman. You will observe in the case of the lady's rings that the diamonds are transparent, while the metal portion casts a shadow even through the bones. These two photographs were taken by Mr. J. W. Gifford, of Chard, an early and most successful worker with Röntgen rays.

Fig. 151 is the hand of Lord Kelvin, and shows traces of age, and of a tendency to rheumatic deposits.

Fig. 152 is the hand of Mr. Crookes, and though a knottier hand, shows some points of resemblance with that of Lord Kelvin.

Fig. 153 is the hand of Sir Richard Webster. The shadow is interesting as showing not only an athletic development, but as revealing, embedded in the flesh between the thumb and the first finger, two small shot, the result of a gunshot wound received many years previously. This photograph and the two preceding are from the series taken by Mr. Campbell-Swinton, who was first in England to put into practice this newest of the black arts.

By the courtesy of Mr. Campbell-Swinton I am also able to show you a number of other slides—the hand of Lord Rayleigh; the hand of a lady with a needle

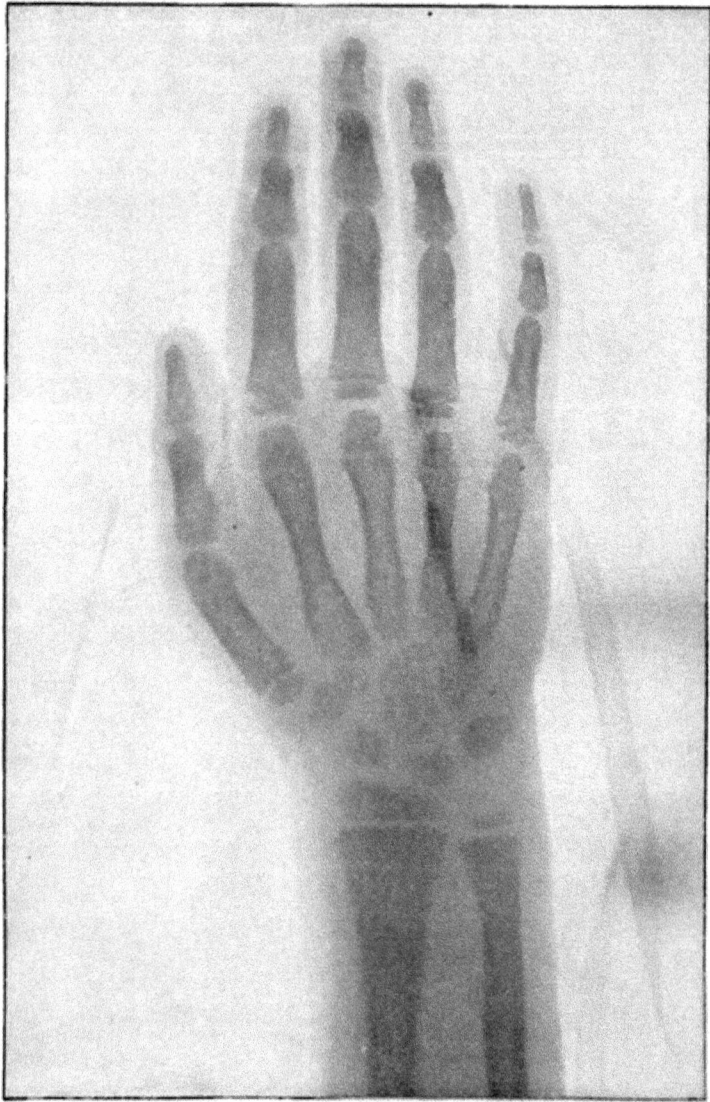

FIG. 149.—Hand of Child, aged eleven years.
(Photo. by Mr. J. W. Gifford).

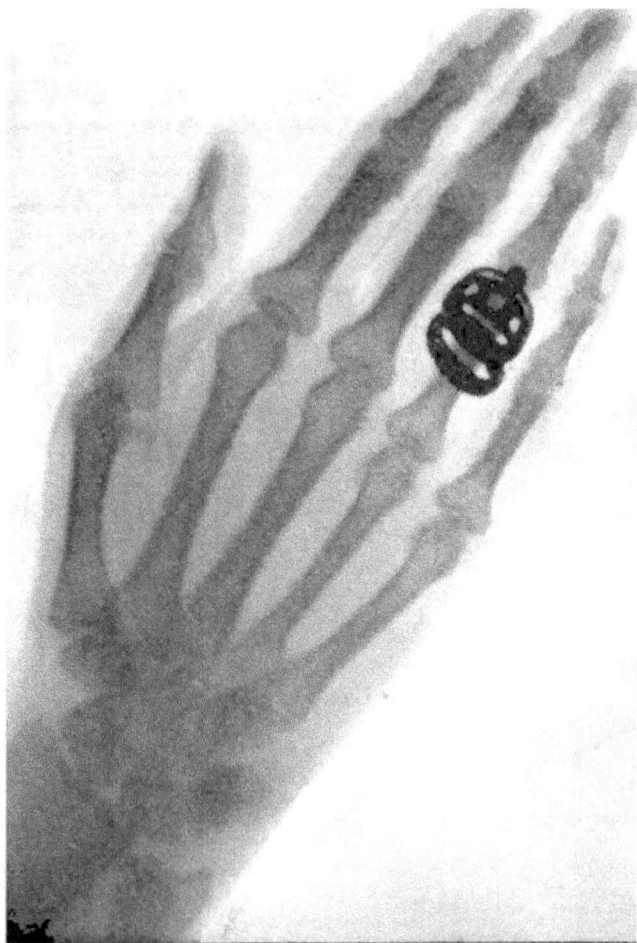

FIG. 150.—Hand of full-grown Woman.

(Photo. by Mr. J. W. Gifford).

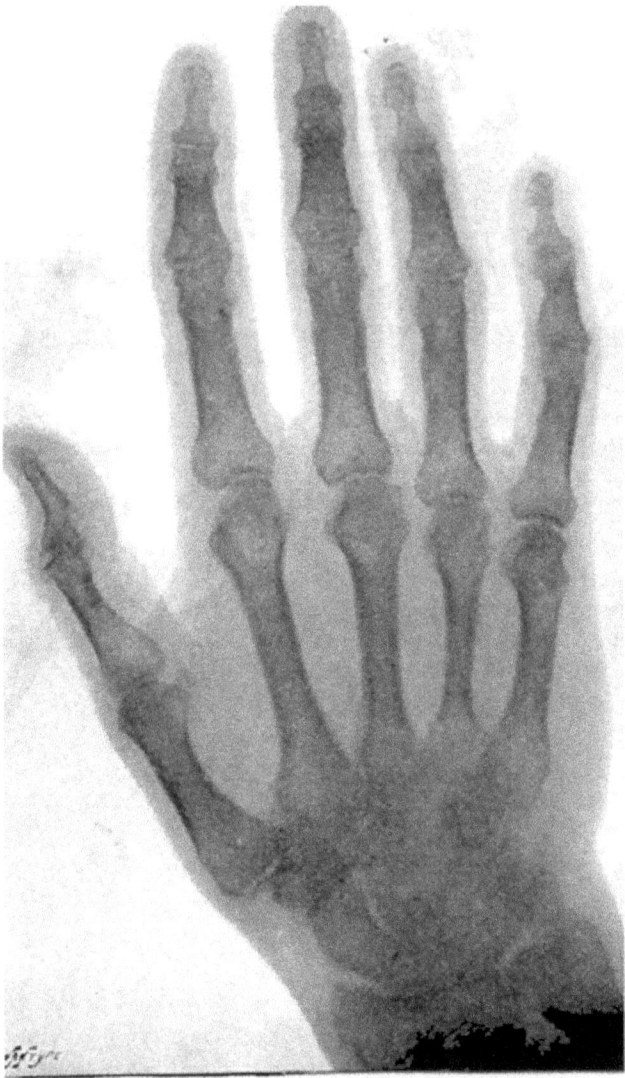

Fig. 151.—Hand of Professor Rt. Hon. Lord Kelvin.

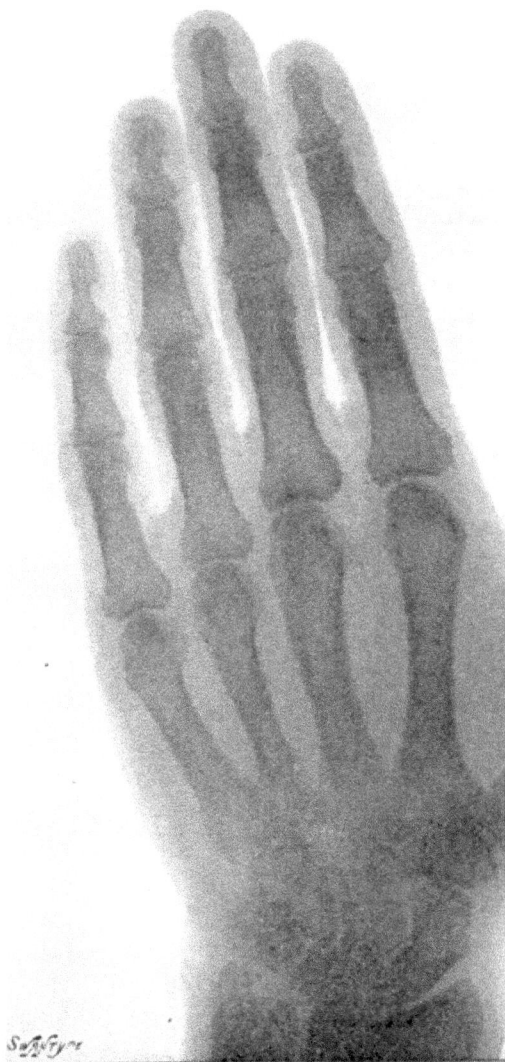

Fig. 152.—Hand of Professor W. Crookes,

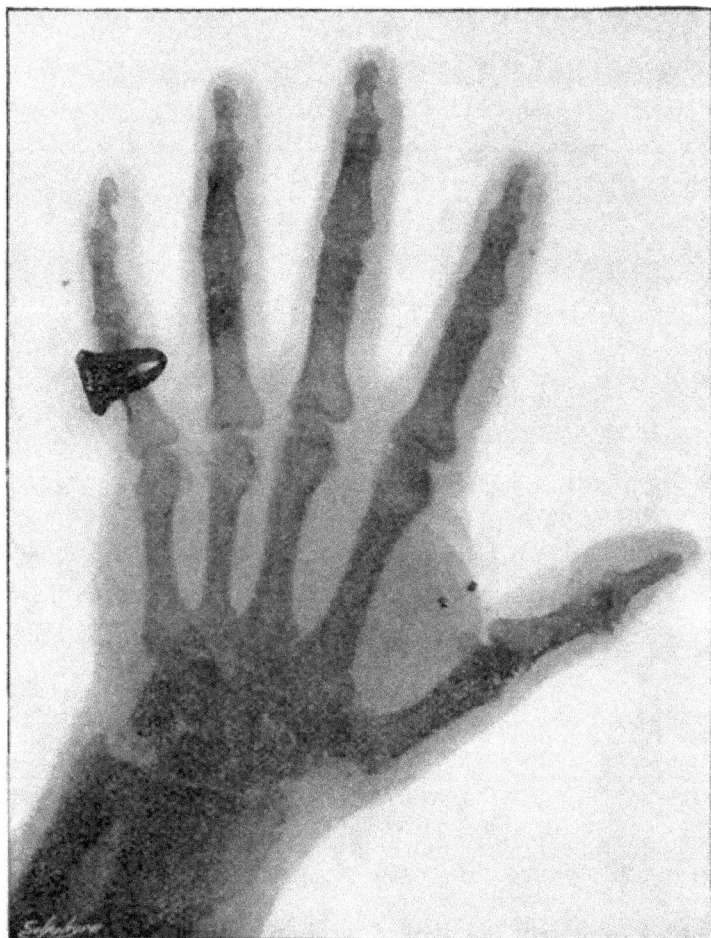

Fig 153.—Hand of Rt. Hon. Sir Richard Webster, M.P.

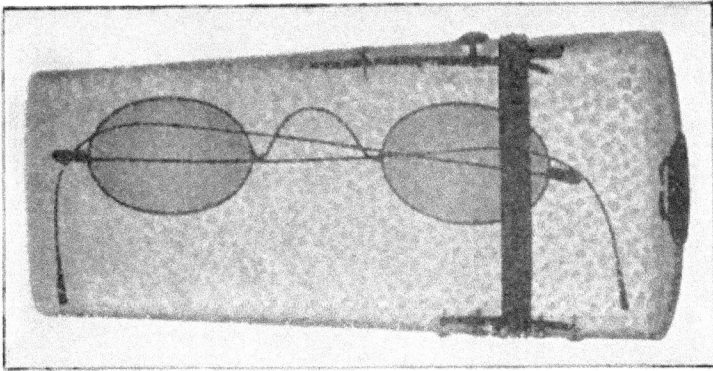

FIG. 154.

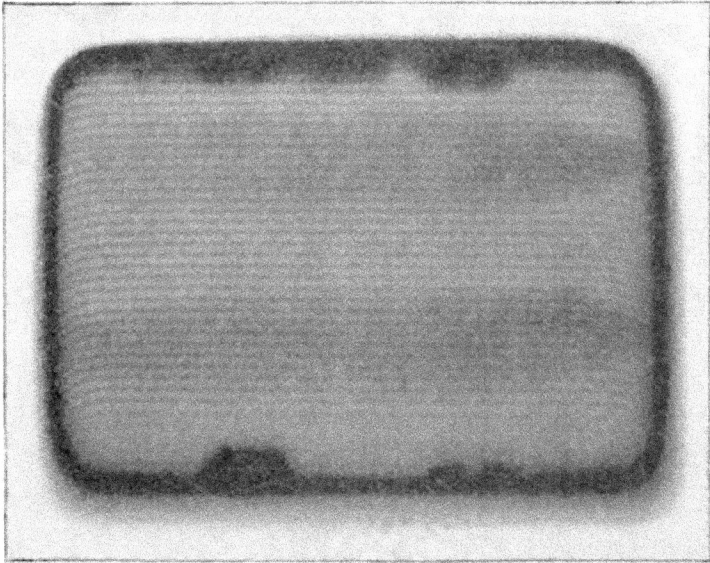

FIG. 155.

embedded in the palm ; a hand terribly swollen with the
gout ; a foot, showing the heel-bone and the smaller
bones down to the toes, as well as the bones in the
ankle ; a view through the left shoulder of a young lady,
showing her ribs, shoulder-blade, and collar-bone ; the
torso of a young man, showing his ribs, and, dimly, his
heart, like a central dark shadow with a triangular apex
pointing down toward the right, that is, to his left side ;
lastly, the shadow of a living head, showing all the
vertebræ of the neck.

Here, again, is the shadow of a newly-born child,
taken by Mr. Sydney Rowland.   Note the imperfect
state of the bones in the hands.

Passing from human objects, we´ will look at the
shadows of a few animals.   These are a chameleon,
giving a clear view not only of its skeleton but of the
internal organs ; a mouse ; a frog ; and some fishes.
The next slide was taken from an Egyptian mummy in
its wrappings.   Before this photograph was taken there
was some dispute as to whether it was the mummy of a
cat or of a girl.   The photograph sets the question
entirely at rest.

Earlier in my lecture I mentioned that glass is tolerably
opaque to these rays.   Of this you have a proof in the
next photograph (Fig. 154), which represents a pair of
spectacles photographed while lying in their case, the
covering of which, in shagreen, shows all the markings
peculiar to the shark's skin, with which the case was
covered.   The next photograph by Mr. Campbell-
Swinton enables you to read the contents of a sealed
letter which he received.   His also is the next picture

(Fig. 155), which is the photographed shadow of an aluminium cigar-case, containing two cigars. And

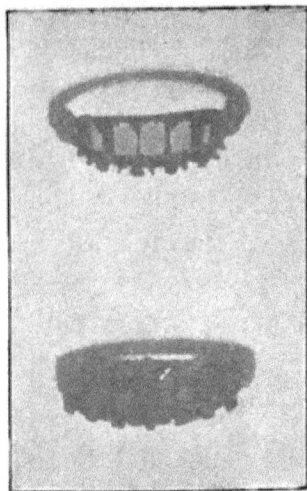

FIG. 156.

lastly (Fig. 156), I exhibit to you a photograph of two ruby rings. By gaslight the gems of one are not distinguishable from those of the other; and in broad daylight it would take an expert to pronounce between them. But when viewed or photographed by Röntgen light there remains no manner of doubt. The rubies of one ring are true Burmese rubies, and they appear transparent. The others are imitation rubies made of ruby-coloured glass, and appear quite opaque.

You will have noticed that I have spoken of these rays as "Röntgen light." But are we really justified in calling it light? It is invisible to our eyes; but then so also is ordinary ultra-violet light, and so is infra-red light, and Hertzian light. And there are other kinds of light too, amongst them one discovered during last year by M. Becquerel [1] and myself, which are invisible. But if

[1] M. Henri Becquerel (see *Comptes Rendus*, cxxii. pp. 559, 790, etc.) and I myself (see *Philosophical Magazine*, July, 1896, p. 103) quite independently discovered some invisible radiations that are emitted by uranium salts, and by the metal uranium, which can affect photographic plates, and will pass through a sheet of aluminium or of cardboard. But they are not the same as Röntgen rays, since, as M. Becquerel has shown, they can be reflected, refracted, and polarised. They also produce diselectrification.

the Röntgen light can be neither reflected nor refracted,
neither diffracted nor polarised, what reason have we
for calling it light at all ?   In fact, direct proof that it
consists of transverse waves is wanting.   Many conjec-
tures have been formed respecting its nature.   Röntgen
himself suggested that it might consist of longitudinal
vibrations.   Others have suggested ether streams, ether
vortices, or even streams of minute corpuscles.   At one
time the notion that it might be simply an extreme kind
of ultra-violet light of excessively minute wave-length
was favoured by physicists, who were disposed to explain
the absence of refraction, and the high penetrative power
of the rays upon von Helmholtz's theory of anomalous
dispersion, according to which the ultra-violet spectrum
at the extreme end ought to double back on itself.

The most probable suggestion yet made, and the only
one that seems to account for the strange lateral emission
of the rays right up to the plane of the antikathode (see
Fig. 142, p. 265), is that of Sir George Stokes.   Stokes's
view is that while all ordinary light consists of trains [1]
of waves (Fig. 68, p. 112), in which each ripple is one of a
series that gradually dies away, the Röntgen light consists of solitary ripples, each of not more than one or

There can be no question that these rays, which are due to a sort
of invisible phosphorescence, consist of transverse vibrations of a
very high frequency : that is, they are ultra-violet light of a very
high order.

[1] It has long been known from the experiments of Fizeau, that
in ordinary light each train consists on the average of at least
50,000 successive vibrations ; for it is possible to produce interference of light between two parts of a beam which have traversed
lengths differing by more than 50,000 wave-lengths.   Michelson
has gone far beyond that number.   See the footnote on p. 112.

one and a half waves. According to Stokes the Röntgen
light is generated at the antikathode by impact of the
flying negatively-electrified molecules (or atoms) which
constitute the kathode stream. At the moment when
each of these flying molecules strikes against the target
and rebounds, there will be a quiver of its electric
charge ; in other words, the charge on the molecule will
perform an oscillation. Now that electric oscillation
will be executed across the molecule in a direction generally normal to the plane of the target, and will give rise
to an electro-magnetic disturbance which will be propagated as a wave in all directions, except where stopped
by the metal of the target. And this oscillation being
of excessively short period, and dying
out after about one or two (Fig. 157)
complete periods, will generate a wave,
which, though of a frequency as high as,
or even higher than, that of ordinary
ultra-violet light, and therefore capable
of producing kindred effects, will not be
capable of being made to interfere, nor to undergo
regular refraction or reflexion, because it does not
consist of a complete train of waves. Here is a model
intended roughly to illustrate the theory. An iron hoop
(Fig. 158) which can be thrown or swung against the
wall represents the flying molecule. The electric charge
which it carries is typified in the model by a lump of
lead capable of sliding on a transverse wire, and held
centrally by a pair of spiral springs. When this model
molecule is caused to strike against the wall and rebound,
the leaden mass is disturbed, and executes an oscillation

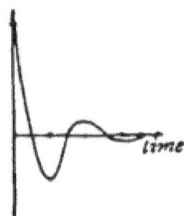

FIG. 157.

to and fro along the wire. The oscillation dies out after
about 1½ periods. Now, suppose this oscillation to set
up a transverse wave in surrounding
space. Though it consists of but
1½ ripples, they would be propagated
outward just as trains of waves are.
And if there were millions of such
flying molecules in operation, these
solitary ripples might come in millions one after the other, but not
regularly spaced out behind one
another like the trains of waves
constituting ordinary light   This is

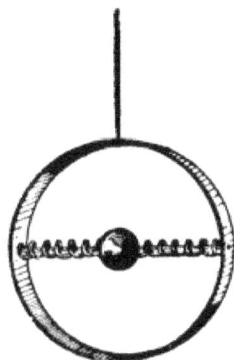

FIG. 158.

but a gross and rough illustration of Stokes's hypo-
thesis; but it must suffice for the present.

But I cannot close this course of lectures without
one word as to the possibilities which this amazing discovery of the Röntgen light has opened out to science.
It is clear that there are more things in heaven and
earth than are sometimes admitted to exist. There are
sounds that our ears have never heard : there is light
that our eyes will never see. And yet of these inaudible,
invisible things discoveries are made from time to time
by the patient labours of the pioneers in science. You
have seen how no scientific discovery ever stands alone :
it is based on those that went before. Behind Röntgen
stands Lenard; behind Lenard, Crookes; behind
Crookes the line of explorers from Boyle and Hauksbee
and Otto von Guericke downwards. We have had
Crookes's tubes in use since 1878, and therefore for
nearly twenty years Röntgen's rays have been in exist-

ence, though no one, until Röntgen observed them on
8th November, 1895, even suspected [1] their presence
or surmised their qualities. And just as these rays
remained for twenty years undiscovered, so even now
there exist, beyond doubt, in the universe, other rays,
other vibrations, of which we have as yet no cognisance.
Yet, as year after year rolls by, one discovery leads to
another. The seemingly useless or trivial observation
made by one worker leads on to a useful observation by
another; and so science advances, "creeping on from
point to point." And so steadily year by year the sum
total of our knowledge increases, and our ignorance is
rolled a little further and further back; and where now
there is darkness, there will be light.

[1] It is but fair to Professor Eilhard Wiedemann to mention that
in August 1895 he described some " discharge-rays " (Entladungsstrahlen) inside a vacuum tube, which, though photographically
active, refused to pass through fluor-spar, and were incapable of
being deflected by a magnet. But their properties differ from
Röntgen rays in some other respects.

APPENDIX TO LECTURE VI

OTHER KINDS OF INVISIBLE LIGHT

UPON the discovery by Röntgen of the rays that bear his name it was natural that the inquiry should be raised whether there exist any other rays having penetrative properties in any degree similar. Lenard's rays, discovered in 1894, to which some reference is made on p. 258 above, have the power of penetrating thin sheets of metal and of producing photographic action as well as of discharging electrified bodies. But they differ from Röntgen's rays in their penetrative power, for air is relatively opaque to them. Also they are deflected in varying degrees by the magnet. Wiedemann's "discharge-rays," briefly metioned above, are further described on p. 281.

No other source than that of the highly-exhausted vacuum tube under electric stimulation has yet been discovered for Röntgen's rays. Many persons have supposed Röntgen's rays to be produced by electric sparks in the open air, simply because such sparks will fog photographic plates and cause images of coins and other metal objects in contact with the plates to impress images upon them. These images are, however, due to direct electric action. They are not produced when a sheet of aluminium is so interposed as to screen off all direct electrical action.

In sunlight there do not appear to be any Röntgen rays, nor yet in the light of the electric arc ; for neither of these sources contains any rays that will affect a photographic plate that is protected by an aluminium sheet.

There are, however, some kinds of light that, like Röntgen's rays, will pass through aluminium or through black cardboard, and produce photographic effects. These are worthy of some notice.

Becquerel's Rays.—Early in 1896 M. Henri Becquerel, as mentioned on p. 272, and the author of this book independently, made the observation that some invisible radiations are emitted from some of the salts of the metal uranium, as, for example, the nitrate of uranyl and the fluoride of uranium and ammonium. These and other salts of uranium, whether in the dark or in the light, emit a sort of invisible light, which can pass through aluminium and produce on a photographic plate shadows of interposed metal objects. This effect appears to be due to an invisible phosphorescence of a persistent sort. Just as luminous paint goes on emitting visible light for many hours after it has been shone upon, so these substances go on month after month emitting an invisible light. Hence the phenomenon is known as one of *hyper-phosphorescence.* It is significant to note that ordinary luminous paint, which ceases after a few days to emit any light of the visible kind, will continue, even for six months, to emit in the darkness an invisible radiation of light that will fog a photographic plate.

Some time after the original discovery, Becquerel observed that metallic uranium far surpasses the salts of that metal in the power of hyper-phosphorescence.

The following is a summary [1] of Becquerel's observations :—A photographic plate was enclosed in a double layer of black paper, over which was placed a thin crust of the transparent crystals of the double sulphate of uranyl and potassium. After several hours of exposure to the sun the plate was found to have been affected. It was also affected when a sheet of aluminium was interposed, but a coin placed between the crystals and the plate cast a photographic shadow. The experiment also succeeds if a sheet of glass is interposed. This double sulphate, when examined

---

[1] See paper by M. Sagnac in *Journal de Physique*, May, 1896, p. 193, and sundry papers by M. Becquerel in the *Comptes Rendus.*

in the phosphorescope, is found to phosphoresce for but $\frac{1}{100}$th sec. after exposure to light. Nevertheless it continues to emit photographically active rays for many hours without being exposed to light. The persistence of these non-luminous rays is incomparably greater than that of the visible luminescence. The invisible rays are enfeebled by passing through either thin glass or sheet aluminium.

Other substances were tried. M. Charles Henry had found sulphide of zinc, and M. Niewenglowski had found sulphide of calcium to emit, under stimulus of Röntgen rays, rays that traversed opaque bodies. M. Troost found hexagonal blende, previously exposed to sunlight, to photograph through cardboard. M. Becquerel found that the special kind of sulphide of calcium which luminesces blue or green gave at first a strong photographic action through aluminium 2 mm. thick. Later the same substances refused to work. Zinc sulphide (hexagonal blende) similarly failed. Becquerel was unable to revivify the sulphides of calcium, either by warming or by cooling to -20° C. and exposing to magnesium light, or by exposing them to electric sparks. In the case of uranyl salts, neither magnesium light nor Crookes tube radiation (to which they are opaque) augment their emission of the invisible rays. Daylight quickens them a little. But the double sulphate of uranyl and potassium is quickened by exposure to the arc-light or to that of sparks from a Leyden jar. Uranous salts, which are not phosphorescent, work as strongly as uranic salts. Crystals of uranium nitrate melted and crystallised in the dark, or dissolved, and which are not fluorescent in the ordinary sense, work just as well as crystals that have been exposed to light.

Becquerel's rays possess, like ultra-violet light and like Röntgen's rays (though to a lesser degree by reason of their lesser intensity), the property of diselectrifying charged bodies. Using an electroscope designed by M. Hurmuzescu, enclosed in metal, he placed a lamina of the crystalline mass of sulphate of uranyl and potassium against an aluminium window 0·12 mm. thick. The leaves, which had an initial divergence of 18°, collapsed in 2 hours 55

minutes. They collapsed in about one-fourth the time when
the lamina was placed inside under the leaves.

These rays are absorbed by air. Water is transparent
to them. Metallic solutions are transparent, as also are
wax and paraffin. Uranium glass and red glass (thickness
2 mm.) are fairly opaque. Native sulphur is transparent ;
calc-spar not very transparent ; quartz is more opaque than
calc-spar. Tin is more opaque than aluminium ; and cobalt
glass more opaque than either. These rays go relatively
more easily through metals than do Röntgen's rays. Copper
(0·1 mm. thick) is very transparent ; platinum nearly as
much. Silver and zinc allow these rays to pass, but lead
0·36 mm. thick is opaque. These measurements were made
electroscopically. Quartz 5 mm. thick is less absorbent
for Becquerel's rays than for Röntgen's rays. Becquerel's
rays will discharge an electroscope through metal screens,
such as platinum or copper 1·4 mm. thick, which would
arrest Röntgen's rays. Copper and aluminium screens are
almost equally transparent. Platinum is a little more
absorbent. The absorption of metallic screens of copper
and aluminium together, or of platinum and aluminium
together, is less than the sum of the absorptions by the
same screens separately, thus proving that the Becquerel
rays are not homogeneous.

Becquerel proved reflexion by laying on a dry-plate a
lamella of uranico-potassium sulphate half-covered with a
polished steel plate, face downwards. The covered part
was more strongly affected after an exposure of 55
hours. He also demonstrated reflexion by using a hemi-
spherical tin mirror. Refraction was demonstrated by using
a thin prism of crown glass, near the refracting edge of
which was placed a tube, 1 mm. in diameter, containing
crystals of nitrate of uranyl. After three days' exposure he
found the light refracted towards the base of the prism.
He proved these rays to be capable of polarisation, since
the photographic shadow through two plates of tourmaline
was stronger when their axes were crossed than when they
were parallel.

Metallic uranium surpasses all other materials in the

freedom with which it gives off these rays. It continues to emit them for months in complete darkness; the source whence this supply of energy is derived being at present unknown.

The author made some experiments to compare the penetrating power of uranium rays with that of Röntgen's rays. Using a layer of uranium nitrate, he found carbon and copal to be as opaque as rock salt; whereas with Röntgen's rays those substances are much more transparent than rock salt.

There appears to be no doubt that the uranium rays are a species of extreme ultra-violet light, having a wave-length certainly less than 10 micro-centimetres, and a frequency certainly greater than 3000 billions per second.

Quite recently Dr. W. J. Russell has found a similar action to be produced by newly cleaned metallic zinc, and by some other materials, including some kinds of wood and paper.

Phosphorus Light.—The author has examined the penetrative effect of some other kinds of light. The pale light emitted by phosphorus when oxidising in moist air is accompanied by some invisible rays which will penetrate through black paper or celluloid, but will not pass through aluminium. So will some invisible rays that are emitted by the flame of bisulphide of carbon.

Light of Glow-worms and Fireflies.—Dr. Dawson Turner has found that the light emitted by glow-worms contains photographic rays which will pass through aluminium.

In Japan, Dr. Muraoka has examined the rays emitted by a firefly ("Johannis-käfer"). He found that they emitted rays which, after filtration through card or through copper plates, would act photographically. These rays can be reflected, and probably refracted and polarised. He used about 1000 fireflies shut up in a shallow box over the screened photographic plate.

Wiedemann's Rays.—Professor E. Wiedemann in 1895 described some rays (named by him Discharge-rays, or *Entladungsstrahlen*) which are produced in vacuum-tubes by the influence of a rapidly-alternating electric discharge. They have the property of exciting in certain chemically prepared substances, notably in calcium sulphate containing a small percentage of manganese sulphate, the power of thermo-luminescence. In other words, the substance after exposure to these rays will emit light when subsequently warmed. They are emitted at lower degrees of rarefaction than are necessary for producing the kathode rays. They are emitted from all parts of the path of the spark-discharge, but more strongly near the kathode. They are propagated in straight lines, but no reflexion of them by solid bodies has yet been observed. They are readily absorbed by certain gases, oxygen and carbonic dioxide, but their production is promoted by hydrogen and nitrogen. Those produced in hydrogen are partially transmitted by quartz and fluor-spar. They are apparently not present in the glow discharge. In vacuo these rays are produced by all parts of the discharge. Under the influence of electric oscillations they are emitted, even in some cases at half an atmosphere of pressure, at the boundary of the rarefied gas and the glass wall, even before any visible light is seen. No deviation of them by the magnet has yet been observable. Those produced at relatively great pressures have in general the power of penetrating bodies according to the inverse ratio of their densities.

New kinds of Kathode Rays.—The author has recently found two new kinds of kathode rays. One of these, termed *parakathodic rays*, is produced when ordinary kathode rays strike upon an anti-kathode, as in the " focus " tubes. If the vacuum is low, there are emitted from the anti-kathode, in nearly equal intensity in all directions, some rays that closely resemble ordinary kathode rays. They can be deflected electrostatically and magnetically, and can cast shadows of objects on the glass walls. If the vacuum is high enough for the production of Röntgen's rays, some parakathodic rays are also produced at the same time. They cause the glass bulb to fluoresce over an obliquely limited region as in Fig. 142, p. 265.

The other kind, termed *diakathodic rays*, is produced

by directing the ordinary kathode rays full upon a piece of
wire-gauze, or upon a spiral of wire, which is itself negatively
electrified. The ordinary kathode rays refuse to pass through
the meshes of the gauze, but instead there passes through
a beam of bluish rays, which differ from kathode rays in that
they are not directly affected by a magnet. These diakathodic
rays can also produce fluorescence of the glass where they
meet the walls of the tube, and can cast shadows of intervening objects ; but the fluorescence is of a different kind,
for ordinary soda glass gives a dark orange fluorescence
instead of its usual golden green tint. This orange fluorescence when examined by the spectroscope shows the
D-lines characteristic of sodium.

Goldstein's Rays.—Herr Goldstein has also described
some rays apparently closely akin to those just mentioned.
If a perforated disk is used as a kathode there are produced
some blue rays which stream back behind the kathode
opposite the apertures.

INDEX

THE END